FEAR, SOCIETY, AND THE POLICE

Fear, Society, and the Police examines elements of fear and how they can be controlled and turned into an effective and proper response in an emergency situation. Readers of this book will be exposed to ways fear can become an uncontrolled emotion, often leading to unnecessary acts of violence, and will examine ways and means of using reasoning to overcome unfounded fear.

The author encourages readers to critically assess circumstances in today's society that have caused fear, unrest, and division between the enforcers of law and the people they are sworn to protect. Providing examples of how violence in society has had an impact on police–community relations, this book examines the many facets of fear from several perspectives, including historical, personal, and institutional. Security management courses concentrate on the "how and why" of security, yet to become an effective professional security specialist it is recommended the practitioner become educated in the nuances of fear. This book presents a look into the how and why of fear, and will relate to security personnel as it does to police officers.

The book brings perspectives based on reality and experience. It will be of interest not only to those who work in law enforcement, but also to students in criminal justice, management and leadership, psychology, and sociology courses. As violence in society escalates, professionalism will require more understanding of fear-based emotions.

Dale L. June has been associated with law enforcement his entire adult life as a county deputy sheriff, a city police officer, and a federal agent. He is currently a criminal justice professor at two universities. His previous books relate to private security and VIP protection.

FEAR, SOCIETY, AND THE POLICE

Dale L. June

Routledge
Taylor & Francis Group
NEW YORK AND LONDON

First published 2020
by Routledge
52 Vanderbilt Avenue, New York, NY 10017

and by Routledge
2 Park Square, Milton Park, Abingdon, Oxon OX14 4RN

Routledge is an imprint of the Taylor & Francis Group, an informa business

Library of Congress Cataloging-in-Publication Data
Names: June, Dale L., author.
Title: Fear, society, and the police / Dale L. June.
Description: Abingdon, Oxon ; New York, NY : Routledge is an imprint of the Taylor & Francis Group, an Informa Business, 2020.
Identifiers: LCCN 2019015183 (print) | LCCN 2019018252 (ebook) | ISBN 9780429283451 (Ebook) |
ISBN 9780367246099 (hbk) | ISBN 9780367246112 (pbk) |
ISBN 9780429283451 (ebk)
Subjects: LCSH: Violence--United States--Prevention. | Crime prevention--United States. | Fear--United States. | Anxiety--United States.
Classification: LCC HV7432 (ebook) | LCC HV7432 .J86 2020 (print) |
DDC 363.201/9--dc23
LC record available at https://lccn.loc.gov/2019015183

ISBN: 978-0-367-24609-9 (hbk)
ISBN: 978-0-367-24611-2 (pbk)
ISBN: 978-0-429-28345-1 (ebk)

Typeset in Bembo
by Taylor & Francis Books
Printed and bound by CPI Group (UK) Ltd, Croydon CR0 4YY

"And now, dear little children, who may this story read,
To idle, silly, flattering words, I pray you ne'er give heed;
Unto an evil counselor close heart, and ear, and eye,
And take a lesson from this tale of the Spider and the Fly."

– The Spider and the Fly (Mary Howitt 1799–1888)

CONTENTS

ABOUT THE AUTHOR

Dale L. June is a former U.S. Secret Service Agent assigned to the Presidential Protective Division at the White House with three presidents plus serving in the Sacramento and San Diego field offices. He is also a former U.S. Army military policeman, a Deputy Sheriff (Shasta County, CA), a City Police Officer (Redding and Sacramento, CA), a U.S. Customs Intelligence Specialist, specializing in Organized Crime and Terrorism, and a Private Investigator and Executive Protection Specialist. He earned his M.A. Degree in Criminal Justice from George Washington University and his B.A. Degree in Public Administration from California State University, Sacramento. He has nearly three decades of experience teaching at college and university levels as an instructor of criminal justice, psychology, sociology, human behavior, history and drug and addiction-related courses. He has written security-related articles for international security magazines in Canada, the United Kingdom, Mexico, New Zealand, and Germany.

Mr. June was inducted into the U.S. Martial Artist's Hall of Fame—Martial Artist of the Year (2005). A former Member of the National Black Belt Association, Tae Kwon Do, he has trained in judo, boxing, free fighting, dim mak kung fu, tae kwon do, police hybrid self-defense and combatives, and Wing Chun Kung Fu, the quick method of self-defense, and combat bayonet fighting. Mr. June sees himself as a "warrior" who loves to read, write, and help those who can't help (or protect) themselves, especially children, grown-ups, and dogs.

Other Books by Mr. June

Introduction to Executive Protection
What They Didn't Teach at the Academy: Topics, Stories, and Reality Beyond the Classroom
Homeland Security and Terrorism: Perspective, Thoughts and Opinion
Protection, Security and Safeguards
The ReEvolution of American Street Gangs

About the Paperback Cover

The web represents "the web of life"; everything is connected. The actions of one thing or person affect many. The barbed wire is a metaphor for how fear imprisons those experiencing its horrible consequences. The blue colors indicate the calming assurance of the presence of a police officer arriving on the scene where help is required.

Fear Is a Scorpion

Everyone has fears. Fear is a scorpion residing in the brain. A scorpion's appearance is fiercely frightening and the venom of its sting is capable of paralyzing its victim. Fear hides in the brain until stimulated, attacking its victim with the potential to paralyze.

Fear Is a Liar!

There have been many instances of police shootings when the officer said, "I was in fear for my life," when the circumstances clearly do not sustain the claim. One instance is very vividly illustrated in a police shooting in the back of a fleeing man on April 4, 2015 in North Charleston, South Carolina.

Even Police Cry

"I'm pretty sure I was pouring my heart out to this kid begging him not to jump, crying intermittently, telling jokes, trying to reason with him, whatever I had to do to keep him on the ledge … and not leaning out towards his death. But the next day at work I was numb. I was empty. I sat in a parking lot and cried for 90 minutes after roll call wondering what the next life or death call I'd get would be."

Shannon Reece, Pasadena, California Police Officer—Retired

ACKNOWLEDGMENTS

This is the place where an author takes time and space to thank and recognize everyone who has in some way contributed to the writer's passion for writing a book that he hopes will find a good reception from readers. To that end I wish to say "thank you for your contributions to this particular effort of mine": There are many others whose story I adapted for the purpose of the book. In no special order because everyone is especially important:

Gavin de Becker, author of *The Gift of Fear,* and President/founder of Gavin de Becker & Associates, for the contribution of his "Media Fear Tactics" essay.

Compton College student Faith Minor (age 19), for her contribution of her reflections on Eminem, and his song *Untouchable,* and fear in the "hood."

Doris Molo, for discussing discrimination and cultural ignorance of police in a racial minority community.

Guadalupe Parra, for his thoughts on fear as a young Latino in a gang neighborhood.

Jill (last name withheld), for sharing her story of childhood abuse.

Joseph Weedman, for sharing his story about being a teenager in prison.

Sarah Hendrickson (a National University Student), for interviewing a ten-year veteran of the LAPD and her report of the interview.

Ryan (last name withheld), who contributed his personal story of fear.

Dr. Bernadett Kucher, and other anonymous reviewers, for their prepublication review for the publisher.

Wing Chun Kung Fu World Grandmaster, William Cheung, for writing the Foreword based on his experiences and knowledge of ways to overcome fear.

Kymlee Murphy, VP of Operations and Firearms & Tactics Instructor for Commonwealth Criminal Justice Academy in Alexandria, Virginia.

Shannon Reece, for recounting her story about police PTSD.

Monica Duperon, for her reaction about fear when she was on a protective assignment in Africa.

Kate Taylor, Editorial Assistant of Taylor & Francis Publishing, and Freelance Copy Editor Sarah Webb for their dogged pursuit of creating the best book possible. Without them you would find typo and grammar errors galore. I learned much from their corrections.

FOREWORD

My father, Cheung Wing Kam, was one of the highest ranked in the Chinese-speaking division of the British Hong Kong Colonial Police from 1935 to 1956. In those times, Chinese could only reach the rank of Sergeant Major and could not be promoted to Inspector or beyond.

Despite this, the Chinese police were relied upon to solve most cases in Hong Kong. During World War II, Cheung Wing Kam joined the Chinese Army, reaching the rank of Colonel. After the war he came back to resume his service with the Hong Kong Police.

After World War II, the South China Sea around Hong Kong was a popular pirate haunt. Even with constant patrol by the police and the Hong Kong Navy, the problem persisted. Many months went by and the authorities could find no trace of the pirates. Early one morning, an informant told Wing Kam that there were pirates hiding in a deserted house located at the top of one of the hills on the outskirts of Hong Kong. Without waiting for backup, Wing Kam rushed to the house. When he kicked down the door and ran inside, he saw seven pirates in the lounge room having a meeting. In those days the Hong Kong police regulation-assigned pistol only held six bullets. He yelled, "My name is Cheung Wing Kam and I am placing you under arrest." The pirates had heard about Cheung Wing Kam's fear-some reputation, and surrendered without a struggle. They were marched down the hill by Cheung to the tram where they then caught the boat to Kowloon, and the bus to Kowloon Police Station. Upon reaching the Police Station, Cheung announced to the desk sergeant that he had caught the pirates. The desk sergeant thought that Cheung was playing a joke, as he and the other police at the station did not believe that Cheung, with his six bullet pistol, could have caught these seven notorious pirates.

In another incident, in the dark of the night, Cheung Wing Kam answered a call to check a deserted factory. His partner went to the back, and Cheung walked in the front entrance, climbed the stairs, and saw a shadowy figure dressed in white. Thinking it was a ghost, Cheung fainted. After this incident, he was the butt of many jokes by his comrades.

When I was growing up and practicing Wing Chun, Hong Kong was littered with Triads. Triad members would try to entice young people, including my friends, to join and then commit crimes for them. If they were caught, because the young members were underage the courts were lenient with them. Many of my young friends encountered threats from the Triads. I stood up for them, and because of this, many times I was ambushed by five or six Triad members. On one occasion they came armed with a poison-tipped blade. As I tried to disarm this attacker, the blade slashed my finger, causing it to swell up and become infected.

A week or so later my friends and I were at a diner. The most popular drink there was icy cold Coca Cola, to which we would add a raw egg. It was an amazing drink! On this occasion I noticed that there were four people outside who seemed to be waiting for us. I also noticed that one of them was holding the poisoned blade that cut my right index finger. So I walked straight up and asked the leader of this group, "Are you looking for me?" "Yes," he said, "look at your finger!" Without a second thought, I jabbed him in the eye with my injured index finger. He screamed and ran away with his friends, holding his eye. It hurt me so much because of the infection, but two days later my finger healed because the pus came out with the jab!

The opposite of fear is courage. Sometimes you do have to do the unexpected to overcome an adversary and control fear. To overcome fear is to confront the thing most feared. Fearing death is common among humans. Yet there are people who face death as recreation; they skydive, swim in shark-infested waters, scale sheer mountain cliffs without the safety net of ropes and equipment. To not control fear is to lead to helplessness and to give up hope.

My strategy to overcome fear or to control it when faced with a challenge is based on *Awareness and Preparedness*. Understand your environment, understand your limitations and understand the situation. The only winner in any confrontation is the one who is able to walk away safely. So awareness means to know your limitations and to know what could possibly happen unexpectedly. Preparedness means to be prepared for specific situations. To be able to use techniques and strategies, you must keep calm and control your adrenaline rush.

Remember in the old days when boxers would come out into the ring and slap the back of their shoulders with their gloves. This is a physical way of stopping the adrenaline getting to the brain. If the adrenaline gets to the brain it creates tunnel vision and the inability to think clearly. Olympic swimmers would tap their solar plexus before competing to stop the adrenaline from rushing to the heart. The advice of doctors and other experts is often to take a big breath, but this is wrong. When you take a big breath in, it will speed up your heart rate by at least 3 percent. The most effective way to stay calm is to short breathe in and long breathe out, like a sigh. So breathe a long breath out, tapping the solar plexus, and slap the shoulders to control your adrenaline rush.

In Traditional Wing Chun Kung Fu, I teach to just watch one point—the opponent's leading elbow. Focusing on just one thing keeps you calm.

William Cheung
(World Grandmaster, World Wing Chun Association)

INTRODUCTION

The purpose of this book is to make a reasoned and fair look at the physiological and psychological effects of fear and to examine the role of fear affecting everyday citizens and a correlation of fear as it relates to police work. "*Cops can get into a state of mind where they're scared to death ....*"[1]

Fear comes in many packages and has many faces, from a minor phobia to panic-driven terror. It is a powerful emotion and reaction to what is perceived as a life-threatening experience. The circumstances of society today have serious consequences in matters of life and death. All the answers will not be found here but this may be an introduction to the psychology of fear and how it affects individuals and society.

The idea is to open the mind of individuals and society to *think* and understand the power of fear. Understanding comes from education and acceptance of what it is that we are fearful of. Fear is a learned concept. Humans are not born with an instinct of fear as animals are. Humans learn fear from their parents, environment, society, and peers.

The psychology (and physiology) of fear explores the emotions caused by fear: stress, anxiety, phobia, and terror, and offers ways to overcome the fear, and flight or fight reflex resulting from an overload of instant adrenalin flowing from the adrenal glands.

The sociological perspective is addressed with examples of how violence in society has illuminated our days with racist and religious overtones, mass shootings, a creation of an anti-gun sentiment, and charges of police brutality and over-reaction with unnecessary, often lethal, force. The role of social and news media and the use and impact of "trigger" words set to alarm or induce the public to believe the world is a very dangerous and frightful place is, perhaps, an overblown orientation of reality.

Students in all the many courses of a university criminal justice program should be afforded an opportunity for learning about the many fearful concerns of a police officer, such as injury, death, failure to properly respond in a dangerous confrontation, anxiety dreams and stress, and an assortment of insecurities including second guessing and "Monday morning quarter-backing." I believe that most textbooks are boring, crammed with fact after fact and detail with significant filler material to place it all in context. I am of the opinion that books should be written for four reasons: (1) to be informative, (2) to be interesting, (3) to entertain, and (4) to make learning fun (as it should be).

Security management courses concentrate on the "how and why" of security. Yet to become a true professional security specialist or police officer, it is highly recommended the student become educated in the nuances of fear. As a practitioner of law enforcement or security there are often times danger resides on one shoulder as fear rides on the other. This book presents a look into the how and why of fear. It will relate to security personnel as it does to police officers.

The book brings perspectives based on reality and experience. Violence in society will continue to escalate and as it does the criteria for professionalism will require more understanding of fear emotions.

I believe policing is more difficult today than in the day when I was a patrol officer, not because of inherent danger but because everything an officer does or says is being recorded to social media by someone with a police grudge. I believe some recordings *must* be seen in their entire context, not just a 20-second clip, usually taken at some midpoint in the confrontation. I also believe policing, considering officer safety, was more dangerous then, but most officers I worked with used violence as a very last resort.

Yes, there were a few officers I called Nazis because of their heavy-handed tactics, especially with people of color. In one instance, while I was still a probationary officer, my training officer and I spotted a young airman sitting drunk on a street curb at two in the morning. He was 18 years old and away from home (Brooklyn) for the first time in his life. He was crying and unable to care for himself or others; therefore, there was no option but to arrest him for his own protection. He was still crying during the booking procedure. The two booking jailers that night were two of the hardest and most racial haters one could imagine.

Also present was the supervising captain, a thoroughly bigoted person. One of the booking officers told the young man to stop crying and "be a man" while slapping him in the face. The captain then stepped in close to the airman and shook a finger within an inch of the boy's face. "You listen to me, Boy, you are a Nigger, always will be! When you look into a mirror, all you'll ever see is a nigger. Now get your ass into that cell!" At that point the jailers grabbed him and roughly shoved him down a few short steps to the cell where he fell to the floor as the cell door was slammed shut.

Since I was still on probation and newly married I didn't interfere. I realized if I did say something that that would be the end of my police career in that city even before it really began. My guilt carried over to the next day. After some consideration and after my last class of the day at college, I drove to the air base and spoke to the airman's commanding officer. I related to him all the events of the previous night. He called the airman into his office to verify what I had said. I was shocked when I saw the young man's face. It was swollen, bruised, and one eye was nearly shut. The captain took one look and said he would not take any military punishment action and thanked me for coming forward. This happened many, many years ago, but is still fresh in my mind.

I believed then, as I still do, that, ideally, police should be remembered for the good they do and the help they provide, not for a violent encounter invoked by the individual coming to their attention. We teach our kids, for their own safety, that the police are our "friends," yet our kids are also taught to fear them—a double-bladed sword. Idealism means "the way things should or ought to be," according to the highest values. However, reality intervenes and what "should be" falls short of "what is." Ideals are rarely obtained. "Ideals are defined in fairness, tolerance, compassion, loyalty, forgiveness, amity, peace, truthfulness, honesty, integrity, social responsibility, and the four cardinal virtues of Prudence (discretion), Justice, Temperance, and Courage."[2]

Ideally, an officer of the criminal justice system ought, or should, be held to all these values. Realistically, however, most, if not all, fall short of the ideal. Maybe not purposely but because the ideal is a belief in the highest order and reality interferes. What is real may not be ideal. What is ideal is rarely real.

The actual grounding of this book is to explore if fear, in many circumstances, is race based, causing police to over-react because of a racial stereotype—fear of getting hurt, fear they may not do the right thing and will be featured on social and news media, fear they are perceived as not courageous by other officers, fear other officers will consider them "not manly" enough for the job, etc., fear of not going home upon completion of their shift, fear the words "End of Watch" will be written after their name.

Other aspects of this book are intended to shine a spotlight on some of the many dangers an officer faces and the damage psychological strain can hold over society and the police. In a large sense this is not a psychology book focused on psychological theories, experiments, or social consciousness, but it does concentrate on the many psychological refrains and mental questions officers may encounter in the people and circumstances they face.

This book examines Anxiety, Fear, Terror, Panic, Dread, and Phobias as a background that a majority of police violence is rooted in: fear and lack of control; fear and lack of control by the officer; and fear and lack of control by the object of the officer's interest. The book looks at fear by average citizens recounting actual moments of fear in their lives. A girl (aged 5) was abandoned, along with her three siblings, with no food. The terrible things she endured included a house fire, sexual molestation by a stepfather, being threatened with a loaded and cocked gun held to her head, and being beaten. What is the fear of a teenage boy going to prison facing a potential life sentence? That is discussed as written by the boy (now a man) in question.

The intent is to show experienced pure fear and how individuals controlled and conquered it … if they did. I also examine fear of the unknown, the randomness of violence, and how the consequences of fear of the unknown can rattle an entire community, even an entire nation.

We have normalized anxiety and fear to the point where we accept it as everyday living. Every worry, stress, fear, and anxiety is exacerbated nearly to the point of social paranoia. The numbers of mass shootings, police shooting and/or beating suspects, and police being shot or injured have reached near historical levels and fill daily news cycles.

In political and governmental bodies during the 1970s' Watergate years, we survived the fear of a national governmental crisis that could jeopardize our democratic way of life. Currently the political and global morass we are in feels more like a nightmare than that constitutional crisis ever was. What is the real truth? Who can we trust? I may be imagining things and seeing them as worse than they are, but I'm afraid of the trend our country is heading toward and of seeing it sink into the dark days of despair, upheaval, and pestilence. Who will become our John and Sarah Connor (the *Terminator* movie series[3])? Can we be saved from Big Pharma, the Machines, Tyranny, fear, and anxiety? Can we be rescued from ignorance, shadowy demons, and self-serving power-driven mad men who thrive on lies and self-aggrandizement? Are we moving toward a dystopian police state or militarization of law enforcement fostered by a government controlled by the rich and powerful?

In the long run I hope to bring attention to the fact that, although police officers display courage in the face of a dangerous situation, they cannot always rein in their fears, and they may react instinctively in a violent manner with a lack of emotional control. A theory I have from observation is that it is not always the first officer at the scene who gets involved in violence, i.e. beating the individual causing the police incident, but it is the third, fourth, or fifth officers, who may have a fear they will be looked upon as being afraid, or not "manly" or "macho," for arriving late to the scene. It has been said in some police departments, "I just want to get my licks in," or an officer has been shamed into "getting his licks in."

It is not the intention of this book to degrade, denigrate, or disrespect police in any way. I have been involved in law enforcement in one manner or another my entire adult life and I hold much respect for officers who follow the concept that they are here to help and do it with dignity and respect for all with whom they come into contact, regardless of the circumstances.

The awful actions of a few police officers unfortunately hurt the image of all law enforcement officers, causing the general public to lose trust and to become fearful of the police. Negative stories about law enforcement often become national news and comments are shared on social media, creating disrespect for police officers across the country. In an unfortunate failure to "see the big picture" when officers are recorded resorting to physical or lethal violence, the public only sees a portion of the event and not the circumstances leading up to it. In a profession that is rewarding, high profile, and dangerous, the public perception of law enforcement is vitally important. The truth is, a majority of law enforcement officers operate ethically and morally, risking their lives in a profession that does not get enough "thank yous."

There is a difference between fear and danger and the difference is addressed in the ensuing chapters. Fear is an emotional result of a danger-causing circumstance in which the victim experiences a life-threatening belief. In this context there is a whole section (Chapter 15) discussing the question "Is it racism or fear?" There have been far too many instances when a white officer claims, after shooting an African-American male, "I was in fear for my life," when there is no realistic way to believe the officer's life was in immediate danger.

I hope to achieve a level that explains the difference among fear, terror, and panic. Terror and panic are when there is a blind emotional and instinctive response to a situation in which an individual has no control over the security and safety of his life in a certain death situation such as knowing his airplane is about to crash to earth in a spiraling fiery explosion and seeing the earth looming closer and closer until that fatal last nanosecond. That is Fear! More so, it is pure panic!

> Let me not pray to be sheltered from dangers but to be fearless in facing them. Let me not beg for the stilling of my pain but for the heart to conquer it. Let me not look for allies in life's battlefield but to my own strength, Let me not crave in anxious fear to be saved but hope for the patience to win my freedom, Grant me that I may not be a coward, feeling your mercy in my success alone; but let me find the grasp of your hand in my failure.[4]

There were times in man's history when fear overrode social consciousness to commit acts on a scale so large, so frightening that, for moments in time, they became an accepted way of life. Regarding inclusion of historical eras of fear (Chapter 5), this was done to illustrate a long history of fear in society caused by political and religious figures ("God fearing"), in other words fear caused by authority. There are two types of laws—legislative and moral (Biblical). They are meant to establish rules for living in a society of diverse peoples. When those laws are grounded in harsh punishment and fear, society often negatively reacts, motivating the administrator to greater draconian punishment, very often resulting in death.

This is not to say the chosen time frames singled out in Chapter 5 are the only or worst times, but are selected merely as reference points. To understand our present, and guide us to the future, we must understand what has gone before. This will also afford access to material reflecting the experiences and past history of some who have experienced fear. It addresses problems associated with fear and violence in society.

In contemporary times we live in a society steeped in fear as witnessed by continuous surveillance, gated communities, fear of random mass shootings, the proliferation of an armed public, and influence of biased journalism plus alleged over-reaction by the police.

I do not think draconian security measures should be implemented, I think that would just make violence and crimes more frequent and people would lose faith in the government to run the country. I believe it will take time, perhaps decades, but winning the hearts and minds of residents in those areas where they are more likely to be victims of crime and violence is key: perhaps through government-funded programs where police officers are able to interact with the community in a positive way and show coming generations that it is not an "US vs. THEM" thing.

Summary

According to an old Roman saying, "It was fear that made the Gods." The primary purpose of this book is to encourage readers to consider today's society and critically assess circumstances that have caused fear, unrest, and division between the enforcers of law and the people they are sworn to protect. The secondary purpose is to examine the many facets of fear from several perspectives including historical, personal, and institutional.

When looking at the "man in the mirror," ask him "What are you afraid of? Why are you afraid? Do you have the courage to face your fears? Is your fear realistic? Do you over-react to your fear to cover it up? Is the thought of hurt, pain, and death always on your mind or is it just lurking there in the back of your consciousness?"

If, in writing this book, I have opened minds to fear-related contemporary social problems, primarily the triangle of society, police, and fear then my effort is a success.

You can't live your life in Fear!

DLJ, Los Angeles, July 2019

Notes

1 www.usnews.com/news/blogs/washington-whispers/2014/12/05/when-it-comes-to-police-brutality-fear-is-also-a-factor. Retrieved August 4, 2015.
2 Ruggiero, Vincent Ryan; *Thinking Critically About Ethical Issues (5th Edn); Mayfield Publishing, 2001; pp. 101–102.*
3 The *Terminator* series is an American science fiction franchise created by James Cameron and Gale Anne Hurd. It encompasses a series of films, comics, novels, and additional media concerning battles between Skynet's synthetic intelligent machine network, and John and Sarah Connor's Resistance forces and the rest of the human race.
4 Rabindranath Tagore; *Fruit Gathering*; quoted in *On Death and Dying* by Elisabeth Kubler-Ross; Collier Books, MacMillan Publishing Co. New York; 1969.

1

FEAR, SOCIETY, AND THE POLICE

In a February 1788 letter to Thomas Jefferson, Thomas Paine discussed the difference between natural law and civil rights with a balance of security. His explanation depicted an example of 20 strangers arriving in a completely uninhabited country. Each claimed his own sovereign land where he could do as he pleased, whenever he pleased; he was his own law. Each was in complete control of his own sovereignty in his own natural rights.

Each sovereign recognized one commonality, the fear of invasion, and power from each other or of the remaining 19. To guard against invasion, a contract or compact treaty joined all together in a union. To make it work every sovereign had to surrender certain rights for the sake of security, in return the compact guarantied certain rights, defined as civil rights.[1] Those rights were included in the U.S. Constitution as the Bill of Rights, our written guarantee against the government taking away personal rights.

How much freedom and liberty must we forfeit in the name of security and "domestic tranquility"[2] caused by "nameless, unreasoning, unjustified fear and terror"? Cameras and monitored surveillance, complex identification systems, security alarms and 24-hours a day monitoring, private police, schools that look like prisons with high fences, walls, and armed guards, restricted movement of persons convicted of certain crimes, fortified and walled planned communities, locked down and restricted entrances to apartment buildings, metal detectors in public buildings. The list grows daily, all because of fear in society. This is far beyond what Thomas Paine declared to Thomas Jefferson. Yes, times have changed, but sacrificing liberty for security is not new.

Fear of the Unknown

"Fear of the unknown." That statement has rung through every person's mind at one time or another.

> Someone once told me that none of us is actually afraid of the dark; we're scared of what it conceals from us. We're afraid of having something with the potential to hurt us standing right before our eyes and not registering it as a threat.
>
> *Lyn Burdick*

Fear of the unknown is what led Man to religion, a belief in finding a way to explain the unknown or unknowable. The key word in the statement "fear of the unknown" is *fear*. There are many ways to combat fear and one of the ways is based on past experiences of fear. Someone who has experienced combat is less likely to be fearful of a potential attacker than someone whose life has been more sheltered. *When I am afraid, I will conquer my fear.*

Experience of the unknown causes fear, which then triggers an emotional response creating a feeling of a loss of security and alarm about the potential of injury or death. *I will make the unknown known.* We human beings have a psychological need and natural right to feel secure and free. President Franklin Roosevelt, in an address to Congress in 1941, said, "We look forward to a world founded upon four essential freedoms: First is the freedom of speech and expression, the second is freedom of every person to worship God in his own way. The third is freedom from want … The fourth is freedom from fear."[3]

In Abraham Maslow's hierarchy of needs, the need for security ranks just above the physiological needs of air, food, sleep, and sex. In a world of seven billion people and unpredictable forces of nature and activities of Man, security and safety are always temporary.

When the feeling of security or the freedom from fear is threatened, we will do anything to get it back. Unfortunately, these actions can have unfavorable consequences. Fear can make us do many things that we would not normally do. Fear is a type of stimulation; for example, to strike out in unreasonable force and rage. Police work requires an assertive personality nearly by definition. Yet fear, if not controlled, can quickly escalate assertiveness into aggression when an officer claims to be "afraid for my life." In these instances, an officer may respond with unreasonable and unnecessary force to take away a person's freedom or life. An increase in aggression has had a great impact on the public and the police; with the police being widely reported as using unnecessary force, including the use of lethal violence. Naturally this has led to much distrust and public aggression against the police and departmental policies.

A long time ago when I was a young police officer in Redding, California, for some reason, long forgotten, a patrol officer attempted to stop a "suspicious" car at about 2:00 a.m. (probably drunk?). The driver didn't stop and a pursuit began. North of town to the South, East to West, and back again, side streets, main street; several cars in the chase, sirens screaming, red lights flashing. This continued for about half an hour until the driver finally stopped. When asked at the police station why he didn't stop, he replied, "I was afraid to, some people were chasing me."

The interesting fact is that no guns were drawn, the man wasn't thrown to the ground and handcuffed spread-eagled on the pavement. He didn't "move his hand toward his waist" and no officer was "afraid for my life." Times have indeed changed. In today's world, the same scenario probably would have played out with the eluding driver being forced to lie face down on the gravelly pavement, spread-eagled and handcuffed with both hands behind his back while being approached by several officers with guns drawn. Any slight movement other than compliance to the officers' commands to "get down and keep your hands where we can see them," could easily place his life in danger.

Instances of police violence and shootings often come from insecurity and feelings of fear by the police officer; fear often generated by the perspective of dealing with a person of color. It may not be racism as often recognized, but fear caused by ignorance and suspicion of the "Other race." A common theme in most police shootings is, "I was in fear for my life" and "He made a movement toward his waistband." Granted, police work is, by definition, dangerous and any incident may and can suddenly erupt into violence and death. Danger lurks in every contact police have, but fear must reside way in the back of their minds and not overtly control their reactions and responses.

"Violence is not power or righteous, it is merely the sign of a limited mind" (Mark James, President of Panther Protection).

Perhaps police are "over-sensitized" to danger in their training. Officer training results in a form of mental conditioning leading to a belief that officer safety is paramount in any engagement with the public. Spending hours and days on a shooting range, the trainer's emphasis of the dangers an officer is bound to face, and simulated scenarios involving guns and other deadly weapons, can cause subtle changes in the thinking and mindset of officers regardless of whether they are newly graduated from the academy or seasoned veterans.

We have to learn to separate fear from actual danger; simply, because an officer is fearful, it doesn't mean another person has to die. If an officer or a member of the public is in immediate danger of severe bodily harm or death, it may warrant a deadly force response by the officer. But fear in the absence of real danger (action versus anxiety) is too low a threshold for the application of deadly force.

During any encounter with the police it seems more people are there to record the incident rather than using a helping hand to get the situation under control. One type of incident that always garners attention is the encounter when the police must use force to effect an arrest. There are a several questions that lead up to the arrest before force is used in most cases. Why is the officer detaining or attempting to detain that person? Why is the officer using force? Why is that person resisting arrest? Was the force that the officer used reasonable and necessary under the totality of the circumstances?

Training

Training must include defenses against bladed weapons and firearms complete with de-escalation and submission methods such as calm voice control, assurance of no one getting hurt, humor, throwing a blanket over or wrapping the person holding the knife and other non-lethal methods. It is an accepted fact that a person with a gun and/or knife presents a dangerous situation and places the officer in a life or death circumstance. It is a highly tense and emotional moment. In those instances, officers must control their fear and show none. Even though they have been trained to use lethal force when necessary, other tactics should be a progression until they employ the last resort of using deadly means.

A common misconception between the public and law enforcement is that all officers are trained fighters. The officers receive hours of combative tactics in the course of their academic training, but not always sufficient to counter someone bigger, stronger, faster, and perhaps better trained. It has been proven over and over again that a fully trained martial artist can be defeated by a street-tough thug. A police officer, though highly trained, may meet that same street-tough thug.

In the "heat of the moment" times when officers find themselves in a physical fight with an aggressive person they must react instantly and efficiently to subdue their attacker before the fight escalates to a level requiring the use of lethal weapons. Officers must rely on their own training and strength. In a crisis situation or when under stress, people will instinctively revert to the way they were trained, but will operate at a high percentage below their "training" efficiency. "Unarmed defensive tactics" are a required talent that officers must be able to call forth under emotional circumstances often driven by fear. Law enforcement officers need a basic combative method that is easy to master and use under stressful circumstances. They must be trained to learn efficient methods of turning a difficult situation to their advantage. They must gain the mental and physical confidence that they can conquer their fear as well as their emotions, and therefore their opponent.

Typically, officers receive approximately 40 hours of training in the subject of defensive tactics. This training is spread over a period of approximately 26 weeks. Training topics covered and performed are searching techniques, hand-cuffing techniques, punching and kicking, less lethal options, weapon retention, and ground fighting. An officer fresh out of the academy does not have significant real life experience when it comes to defending himself or herself. Most often the new, freshly graduated, officer has never been in a serious fight, never taken a blow to the face and surely has never faced an opponent with a real weapon. It is some officers' opinion that too much time is spent on learning penal codes rather than doing practical exercises or scenario-based training. Often citizens and officers refer to the material on their duty belt as "tools." These "tools" are merely pieces of equipment if the officer is not proficient in how and when to use them.

When it comes to use of force, the police are trained to attempt to deescalate the situation first. This is accomplished simply by talking someone down, utilizing good interpersonal skills. This is the common theme when it comes to police interaction with a subject who is less than cooperative. By statutory law, when an officer has probable cause to believe that the person to be arrested has committed a public offense, he or she may use reasonable force to effect the arrest. Reasonable in whose eyes? This is where the public generally gets it wrong. Reasonable is typically based on the "reasonable officer standard"; weighing the totality of the circumstances and opinion from an officer with training and experience.

Many times when officers have to use force it is because of the lack of control over the person they are attempting to arrest. The lack of control comes from an officer uncertain or doubtful in his or her abilities. Officers are taught to use a "command voice," meaning

> the voice must have carrying power, but excessive exertion is unnecessary and harmful. A typical result of trying too hard is the almost unconscious tightening of the neck muscles to force sound out. This produces strain, hoarseness, sore throat, and worst of all, indistinct and jumbled sounds instead of clear commands. Ease is achieved through good posture, proper breathing, correct adjustment of throat and mouth muscles, and confidence.[4]

When in a stressful circumstance, such as facing a non-compliant individual (perhaps wielding a weapon), the officer's voice might rise to a shrill, high, piercing level, displaying a general lack of confidence, uncertainty, and fear. Naturally, this raises the level of fear, confusion, or indecision on the part of the non-complying party. This creates a tense situation that could easily devolve into a violent exchange.

What it boils down to is a ticking time bomb when a seasoned criminal comes in contact with a hesitant officer. Any person that doesn't want to be arrested or contacted will exploit every possible avenue they can to evade, escape, or resist.

An officer must make a quick decision, based on the totality of the circumstances whether or not he or she will use force to effect an arrest. To overcome fear and doubt, an officer may react in an overly aggressive manner, resulting in inappropriate force. The force that is used generally will be broken down into three categories to determine if the force used is reasonable. The seriousness of the offense, the threat to the public, and the threat to the officer.

In the public's eyes, they only see a small glimpse of what is or has taken place. If the public sees an officer using force on an unarmed black male they will instantly paint the officer as a racist. The public does not see or know the entire details of the incident, therefore they develop a biased opinion of that officer, which in turn paints all officers as racists.

FIGURE 1.1 Police officers may react in an overly aggressive manner when trying to overcome fear and doubt
Source: https://commons.wikimedia.org/wiki/File:Police_brutality.svg

There are some instances that are completely unjustified and utterly unreasonable. For instance, the Rodney King[5] beating. While police brutality is not as common as some may think, the ones that make the headlines spark backlash between the communities and law enforcement. There's a ripple effect caused by these few incidents. The effect oftentimes leads to protest. What it should lead to is reform. An in-depth review of police training is required to emphasize areas that need to be addressed. Not only training, but hiring standards need to be raised in some areas where red flags exist. These red flags include anger issues, racial discrimination, decision-making abilities. The higher caliber of officer means the public will receive a better and fairer treatment from officers of the law.

In decision making and considering all the consequences and obligations, a decision maker must consider what is good for the most benefit and the least harm. It is a delicate balance but a professional experienced in dealing with a varied assortment of personalities should have developed interpersonal "people" skills to the extent that he or she will address all issues of "what is good for all." The decisions will be based on obligations but tempered with consideration of all consequences.

Good decision makers, especially professionals, do not "shotgun" their approach. They zero in on it, giving it just the right spin in an original and thoughtful manner. Creative thinking and decisive decision making are as important to students, housewives, and soldiers as they are to judges, lawyers, police officers, politicians, and executives. In many circles, decision making is portrayed as "thinking on your feet," being flexible enough in the mind to instantly create an answer to a problem, a solution to a predicament, or to defend one's position.

Decision making is not a precise science, especially in a crisis mode, nor is it really an art. It does not depend on a particular talent except being able to anticipate the effects of an action taken or not taken. Often it takes moral courage to make a decision and to abide by it regardless of the fallout or of what others may think. But life is a matter of decisions, and the professional is expected to make decisions that could have life or death implications. The professional measures the consequences of his or her action or inaction and acts accordingly to "do the right thing."

Officers frequently must make decisions measured in thousandths of a second and have no time to consider the consequences of their immediate action. To hesitate and think a situation through to its final conclusion may result in injury or death. Decisions made in those fleeting heartbeats may take attorneys and judges years to unwind and resolve.

Studies of stress have shown that when an officer is faced with a situation that triggers the fight or flight response of an infusion of adrenalin, this will oftentimes lead to tunnel vision. Tunnel vision is the reaction of the brain to narrow focus. This will create a possibly lethal encounter by missing the entire picture and affecting decision-making abilities. Newer officers are faced with so many encounters that are difficult and completely foreign to them. There is no book they can read and there is no field training officer or academic setting that can prepare someone for every encounter. The experience will come with time and involvement. Officers must, at all costs, use the best and most practical resources available to them at that particular time. What it all boils down to is maturity and training. Training is the key to success. You cannot expect a football team to succeed without practice, just like you cannot expect a police officer to succeed without the proper training.

> [A retired veteran Boston police officer said] "I, in no way had the requisite maturity and wisdom," ... "I had never held a gun before. It's a dirty little secret that we're hiring police officers too young." ... "I was in over my head," he said about his start. "The tendency for someone that is overwhelmed or fearful is to react with excessive force, and I say that from personal experience."
>
> The age issue is not limited to a lack of experience on the streets. Experts say the problem is deeper—it's biological. Most people have not matured and developed their sense of judgment fully until their thirties, and police officers sometimes have to make a series of very complicated decisions in a short period of time.
>
> As for the officers in recent high-profile cases, most who pulled the trigger were white. Many had disciplinary records; a number of them had another thing in common: They were under the age of 30. In Ferguson, Missouri, police officer Darren Wilson was 28 when he fatally shot Brown. A month earlier, Eric Garner was put in a fatal chokehold by NYPD Officer Daniel Pantaleo, who was 29 at the time. In Cleveland, 12-year-old Tamir Rice was shot in a playground after Officer Timothy Loehmann, who was then 26, mistook his pellet gun for a real weapon. And in Brooklyn, Officer Peter Liang, 27, who was on the job for less than 18 months, accidentally fired his gun in a public housing stairway, killing Akai Gurley.
>
> "We all recognize that people in their twenties don't have the maturity of judgment that people in later decades do," said Dr. Jeffrey Jensen Arnett, a research professor in the department of psychology at Clark University. He added that occupations that come with heavy responsibility, such as CEOs, surgeons, and doctors are typically not held by people younger than 30 or even 40 years old. "With certain professions more maturity is required, and I would think police officers is one of them. You have the authority of the state behind you—the power of life and death," he said.[6]

Scientific studies have shown that the human brain (prefrontal cortex) is not fully developed until the human is well into the mid-twenties. A consensus of neuroscientists agree that brain development likely persists until *at least* the mid-twenties—possibly until the thirties.

> The prefrontal cortex (PFC) is the cerebral cortex covering the front part of the frontal lobe. This brain region has been implicated in planning complex cognitive behavior, personality expression, decision making, and moderating social behavior. The basic activity of this brain region is considered to be orchestration of thoughts and actions in accordance with internal goals.[7]

There are a variety of functions for which the prefrontal cortex is responsible. A few of the most relevant to the discussion of police maturity and actions are:

1. **Decision making**: As we enter our twenties, our decision making improves. This is due to the fact that our prefrontal cortex helps us think logically and make more calculated assessments of situations. Our brain weighs the risks and tells us whether a certain behavior or choice is a good idea vs. a bad one.
2. **Impulse control**: Struggling with impulsivity is often related to deficits in the prefrontal cortex. The ability to maintain self-discipline and avoid impulsive behaviors has not reached its peak until the twenties.
3. **Logical thinking**: Justifying behaviors based on emotions rather than logic. When the prefrontal cortex fully develops, logical thinking simultaneously improves. This means you will be better at rationalizing and making smarter choices.
4. **Risk management**: The ability to assess risky situations and determine whether they will result in long-term benefit is a byproduct of your prefrontal cortex.[8]

Behaviors when under the influence of fear, especially a police officer's reaction that may or may not be justified, whatever the cause of fear, often define the character of the officer as well as society.

Summary

Danger rides with police officers. Unconsciously, fear and anxiety are also partners. Officer academy training emphasizes semi-military physical fitness, use of firearms, "defensive combatives," and an array of academic topics like study of laws, procedures of arrest, search and seizures, report writing, etc. The normal starting age is 21 but science of neuropsychology has concluded that the human brain, especially the prefrontal cortex, may not be fully developed, restricting the individual's decision-making capability. Questions have surfaced about raising the minimum age level to 25, allowing for maturity and development of the police recruit. The older age should foster greater decision making and confidence in the officer.

Review Questions

1. Explain fully if you agree or disagree with this statement: "Instances of police violence and shootings often come from insecurity and feelings of fear by the police officer."
2. Explain the difference between fear and actual danger.
3. Are police officers over-sensitized to danger in their training in an academy?
4. Are police officers recruited at too young an age?
5. The Bill of Rights and democratically inspired legal restraints on the police help ensure personal freedoms and security in our society and inspire "freedom from fear," so why are many people still fearful of the police?

Notes

1 *Thomas Paine: Collected Writings*; The Library of America; 1995.
2 "Domestic Tranquility" is one of the reasons for the writing of the Preamble of the U.S. Constitution.
3 Quoted in *Understanding Violence and Victimization; 6th Edn* by Robert Meadows; Pearson, 2014.

4 www.armystudyguide.com/content/army_board_study_guide_topics/drill_and_ceremony/the-command-voice.shtml. Retrieved December 29, 2018.

5 On March 3, 1991, King—who was on parole for robbery—led police on a high-speed chase through Los Angeles; later, he was charged with driving under the influence. When police finally stopped him, King was ordered out of the car. Los Angeles Police Department officers then kicked him repeatedly and beat him with batons for a reported 15 minutes. A video taken from an apartment balcony showed that more than a dozen cops stood by, watching and commenting on the beating. King's injuries resulted in skull fractures, broken bones and teeth, and permanent brain damage. www.npr.org/2017/04/26/524744989/when-la-erupted-in-anger-a-look-back-at-the-rodney-king-riots. Retrieved 7 September, 2018.

6 Georgantopoulos, Mary Ann; *Younger Police Officers Are More Likely to Shoot People than Older Ones*; www.buzzfeed.com/maryanngeorgantopoulos/age-of-police-officers-who-kill-people-a-little-discussed-fa?utm_term=.sd0mmypJ1K#.jgbddneoNM. Retrieved May 5, 2018.

7 www.neuropsychotherapist.com/prefrontal-cortex/. Retrieved September 7, 2018.

8 *At What Age Is The Brain Fully Developed*? https://mentalheathdaily.com/2015/02/18/at-what-age-is-the-brain-fully-developed/. Retrieved December 28, 2018.

2

INTRODUCTION TO FEAR

"What Are YOU Afraid of …?"

> **FEAR**—"an unpleasant emotion caused by the belief that someone or something is dangerous, likely to cause pain, or a threat."[1]

Fear—The Ultimate Unknown Factor

Fear has been defined as "an emotional reaction to a perceived or actual life or death threat often resulting in irrational actions."[2]

> Fear is the emotional reaction to an actual and specific source of danger. A survival mechanism, fear is usually related to an apprehension toward pain. Severe fear is a

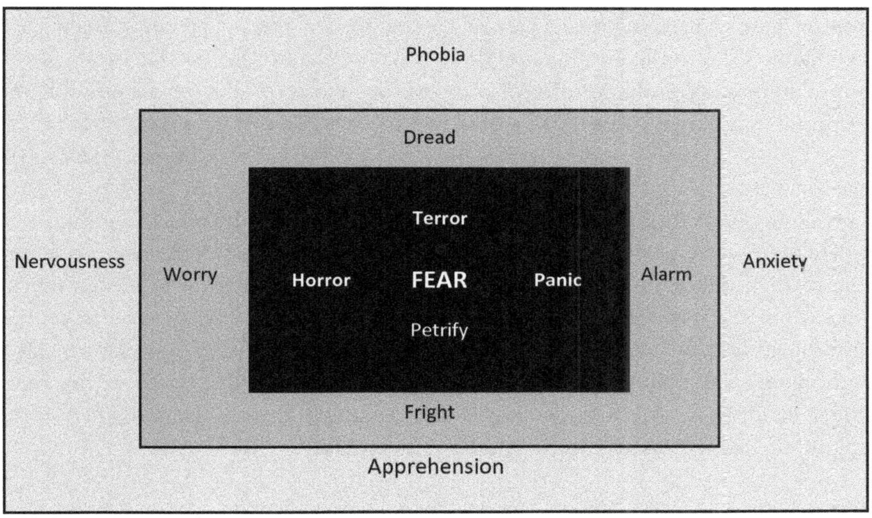

FIGURE 2.1 Elements of fear

> reaction to dreadful approaching danger, and trivial fear occurs as a result of a confrontation that does not pose a significant threat. The degrees of fear vary from slight caution to paranoia. Fear can affect the unconscious mind through nightmares.[3]

Fear is when you dread the consequences of having something to lose. A sense of fear comes with anguish, defensiveness, and isolation, causing overall negative outcomes. To most observers, courage is the opposite of fear. That could easily be understood. In many circumstances that is true. Another realistic definition: safety is the opposite of fear, people want, and need, to feel secure.

Fear is routinely defined as a distressing emotion aroused by impending danger, evil, pain, etc., whether the threat is real or imagined. Fear is living in the future with anticipation of the possibility that something unpleasant will occur.

Fear is a natural instinct; a defense mechanism, both physical and emotional. Fear is triggered when the brain recognizes immediate bodily and life-threatening danger. It is the instinctive reaction by which means a person or animal avoids grave injury and/or death.

A debatable question amongst psychologists is: What is the difference between emotion and instinct?

> Although emotions and instincts appear to be different, they are not. Instincts are automatic emotions. Take the example of fear. Would you consider fear an emotion or an instinct? You can rationally have the emotion of fear but sometimes one experiences fear immediately as a consequence of an event, in an instinctive fashion. ... The difference between emotions and instincts is that one can be rational in emotions while the instincts are immediate, irrational, innate and exist in many different species.[4]

Fear Explained

To capture the essence of fear, it must be understood that each person reacts differently to the cause of fear, the consequences, and degrees.

> One great enemy we must all endeavor to fear not conquering is fear. Fear can cripple purpose and purposeful life. Fear asks question we must fear. Fear makes vision a nightmare. One must always cross the barrier of fear to get to the great city of true purposefulness. A great number of us who are unable to live to accomplish the true reason for our existence on earth are unable to cross the barrier of fear in the first place.
>
> *Ernest Agyemang Yeboah*[5]

Some psychologists define fear as "a motivational state aroused by specific stimuli that give rise to defensive behavior or escape.[6] In more common terms that definition is referred to as the 'fight or flight reflex.' A cornered animal will fight to the last sliver of breath, but he may have fled as his first reaction to the perceived threat to his life or he may be frozen in place hoping to blend into the background."[7] Thus it is in humans. With the physiological changes within the body, a human may initially react with a startle response as his or her body prepares for flight or fight. All of this occurs within a hundredth of a second. There is no time for thinking or rational reaction; all response is involuntary.

> Fear is the emotional reaction to an actual and specific source of danger. Fear is a reaction to an actual danger signal—it involves physical and mental tension that helps you

> spring into action to protect yourself from something that is happening. A survival mechanism, fear is usually related to an apprehension toward pain. Severe fear is a reaction to a dreadful approaching danger, and trivial fear occurs as a result of a confrontation that does not pose a significant threat. The degrees of fear vary from slight caution to paranoia. Fear can affect the unconscious mind through nightmares.[8]

Courage is not the absence of fear, but rather making progress in its presence. "Fear is an aid to the warrior, it is a small fire burning. It heats the muscles making us stronger. Panic comes when the fire is out of control, consuming all courage and pride" (David Gemmell[9]). Fear must not be in control otherwise it gets worse. Fear wants to isolate the fearful, and then has power to attack, "it involves physical and mental tension that helps you spring into action to protect yourself from something that is happening. A fear reaction happens whenever we sense danger or when we are confronted with something new or unknown that seems potentially dangerous."[10]

Fear has conquered many men. The horror, the feeling of helplessness, the fear, the ... *terror* ... *the* ... *panic!* There is no greater weapon than fear. Fear weakens and disables. Fear saps and takes away freewill. For choice to be possible, we must be free to act. Freewill means not to be compelled to act by an irresistible force or emotion either from within or externally. Terror eliminates freewill and physiologically causes changes in the brain and body leading to dizzying blind panic. One of the best ways to conquer fear is to learn it, understand it, and face it. Terror, on the other hand, is fear on super gigantic steroids. It cannot be conquered, understood, or faced. There is no reasoning with "bone chilling terror." The word *terror* comes from the Latin word, *terrere*, meaning "to cause to tremble." The word came into common usage during the *Reign of Terror* in the French Revolution (see Chapter 5).

Anatomy of Fear

Fear is normal. When the normalcy of fear is realized, it can be overcome.

> When you are afraid, you are likely to have:
>
> - worried *thoughts*;
> - physical *sensations* like a faster heart rate, sweating, increased breathing;
> - *behaviors*, like trying to escape the situation that made you afraid in the first place!
>
> There is a complicated series of events in the nervous system that leads to the physical sensations and behaviors of fear. One of these physical events associated with fear is often called the "fight or flight reaction": increased heart rate and force of each beat ("pounding heart"); increased muscle tension that can even cause tremors; sweaty but cold palms; and even nausea and diarrhea. Another aspect of fear is "conditioning," so that even a minor stimulus can bring on the whole fear reaction. The brain structure which appears to be at the very center of most of the brain events associated with fear is the "amygdala" (Greek for "almond," its shape). The amygdala seems to respond to severe traumas with a difficult-to-erase fear response (post-traumatic stress disorder, PTSD). It seems to be genetically different and "wired" for a higher level of fear in some individuals, such as those with panic disorder.[11]

It is a theory that many police violence incidents are a result of over-reaction to fear on the part of the officer. Elements of fear and how it must be controlled and turned into an effective and proper response in an emergency situation need to be examined. The ways fear can become an uncontrolled emotion often lead to unnecessary acts of violence. Controlled emotions and reasoning are paths to overcoming unfounded fear.

Fear has a long history from the darkest moments of cave dwellers to contemporary societies. It is necessary to take into account historical social and religious aspects. One way to look at religious fear in a whole society is by comparing Puritanical law with Sharia law (Chapter 5), both of which call for extreme, and more often, lethal punishment for various and sundry crimes that in the context of today's society are not in the mainstream, but as a whole, bring fear to an entire population under the control of rulers who demand respect and fellowship through fear and intimidation. Fear in a population results in distrust, governmental dystopia, and chaos.

Today's fearful society includes often unnecessary violence from the ones who have sworn to protect us from much of the violence … the police. Incidences of attacks by and against the police, politically, verbally, and violently physical, could have been prevented if threatening indicators and procedures had been addressed. Fear is a loss of control. It causes rational people to lose cognitive thinking, often driving them to acts they wouldn't otherwise commit.

We Are Prone to Fear

> "If we are honest with ourselves, most of us will have to admit that we live out our lives in an ocean of fear."
>
> *Jon Kabat-Zinn*[12]

> We are prone to fear for a reason. Fear of danger can keep you alive. But fear is a fickle beast that can distort and damage as much as protect. Fear prevents exploration, undermines growth and can poison society. Fear can be a great motivator, but also can be a great destroyer. Leadership decisions based on fear among a people seized by fear can lead to the decline and fall of nations. This has been known since at least the time of the ancient Egyptians and Greeks.[13]

In the words of Albert Einstein: "Education is not the learning of facts; It is rather training the mind to think." Training the mind to accept dangerous circumstances and still react courageously doesn't, or shouldn't, include taking time to think of the consequences, but to respond spontaneously. Fear slows the mind. Though an emotion, it is a concept that can be conquered if it is recognized, isolated, and suppressed. Fear of the unknown is the greatest fear of all. We associate death with fear, or the circumstances causing the fear.

Have you ever been at home alone at night and you sit in quiet comfort when suddenly you hear a noise that startles you because you don't recognize it as the house simply settling or creaking? You imagine all potential sinister scenarios. To allay the fear, you turn on the television just to assure you the noise is natural and there is nothing to be afraid of. This is very common with people who are unaccustomed to living alone. The television, radio, or CD music player covers the sound, or imagined sound, thus creating a climate of not hearing a strange noise, ensuring the feeling of security, though it may be a false sense of security. It becomes an attitude of "If you can't hear a strange noise, it doesn't exist."

English writer Charles Dickens, known for some of his dark stories of ghosts and haunting, captured some of that feeling of the unknown in his short story *The Haunted House*:

> In the dead of night try this at your own comfortable fire-side, in the life of the night. You can fill any house with noises, if you will, until you have a noise for every nerve in your nervous system.[14]

Stress, anxiety, apprehension, and concern are merely minor players when compared with the hidden and dark, internalized emotional makeup of law enforcement officers, firefighters, soldiers, and many others who know that when they leave home in the morning, possibly the only way they will return is in the back of a coroner's wagon. Post-traumatic stress disorder; Serious emotional prices to be paid? Yes. Yet, there is another emotion that trumps all these others and is seldom, if ever, discussed by those brave men and women—fear.

Much has been said about the role of police and society in violent police encounters with the public. By definition, there is a schism between police and the people they are attempting to provide law enforcement services for. No one enjoys a police encounter unless officers are responding to a call for help, but even then the circumstances may be less than favorable. If a police contact is initiated without a request by the person being stopped for a traffic violation or because of suspicious circumstances, a certain amount of fear, anxiety, and nervousness can be expected from the individual. Officers must realize the emotions their mere presence provokes.

In an incident occurring in Ferguson, Missouri during an observance of the one-year anniversary of a white police officer shooting and killing a black man, an 11-year-old girl approached a police officer and told him, "I don't like you. I am afraid of you!"

The attitude and demeanor of the officer play huge roles in soothing the fear and anxiety of children, teenagers, and adults. To be feared or to be respected is a genuine question. Of course an officer expects to be respected, not feared. Respect of police is earned, fear of police is a result of many factors like "heavy-handed" tactics, poor interpersonal skills, and perhaps an officer over-reacting to disguise his or her own fear. With a sense of fear comes defensiveness, isolation, and overall negative reactions. Many people in power positions mistakenly endorse fear; misidentifying it for respect. Sometimes fear is generated by overblown egos of those in control who enjoy having power and manipulation over others. The

FIGURE 2.2 Would you feel fear if you were alone in this house at night?
Source: Author's own photo.

thin line between fear and respect develops because of variables like time, place, and circumstances, and the individuals affected. A common management or leadership philosophy is "fear is a great motivator." Fear of police has many negative consequences and respect lends positive connotations.

Fear of Random Violence

"So first of all let me assert my firm belief that the only thing we have to fear is fear itself—nameless, unreasoning, unjustified terror which paralyzes needed efforts to convert retreat into advance." In his first inaugural address, President Franklin Roosevelt was, of course, addressing the economic conditions of the Great Depression. But now let us look at issues of the modern day that FDR could also be referring to ... fear of random violence, mass killings as gunmen attack educational institutions (school shootings), churches, airports; police officers being attacked and shot, or fatal shootings by police, and the potential threat of aggressive terrorist attacks.

In March 2018 the whole city of Austin, Texas was in fear and the country felt reverberations of terror from the placing and detonations of six bombs by an unknown bomb maker, reminiscent of the "Mad Bomber"[15] of New York in the 1940s and 1950s and of the Unabomber Theodore Kaczynski who killed three people and wounded 23 others with mailed letter bombs between 1976 and 1995.

In Austin, the unknown bomber (later identified as Mark Anthony Conditt who confessed in a video recorded on his phone prior to blowing himself up as the police closed in on him[16]) used homemade explosive material and placed them in packages to be left at random places. In late October 2018, only a week before the national midterm political elections, pipe bombs were mailed by an unknown person of interest to several democratic politicians, including two former presidents, a former CIA director, a state governor, and Hollywood celebrities who supported democrat party platforms. Fear began to arise among democrat party supporters. When the dark specter of an unknown bomber, a serial rapist, or killer is raised in a community, fear has citizens asking, "Who will be next?" "When will he strike again?" "Is anybody safe?"

Creating fear in a victim is a major factor in the psyche of a serial killer, who will select his prey, stalk, capture and incapacitate his victim; teasing with threats and watching the victim struggle as the fear grows and grows, evident in the victim's eyes, facial expressions, and body movements. Getting tired of "playing," he will finally make that fatal leap! An ancient Colombian word, *Yaguara*, means, "He who kills with one leap."

The 1960s came to be known as "The Decade of Violence" because of political assassinations, city riots, anti-war demonstrations, civil disobedience, and labor strikes. Police were called "pigs," "fuzz," and other disrespectful names; the streets became combat zones as riots challenged the police and neighborhoods burned. The consequential effects on the American public had long-range significant tremors, still being felt today.

The Constitution's First Amendment guarantees freedom of assembly (protest demonstrations, for example) and limits the right of the people to "peaceful assembly." Today's world of public "demonstrations" has spun way beyond the intended "peaceful assembly," with many such marches and public gatherings to protest a governmental action turning violent, resulting in injury and death. Police reaction to crowd activity could lead to greater violence. At the 1968 Democratic National Convention, police over-reacted to unruly crowd movement, resulting in massive police violence in what has come to be regarded unfavorably as a "police riot."

Police officers are cognizant of the dangers they face, yet the fear of those dangers must be pushed out of their consciousness, for fear makes people act irresponsibility and irrationally. Fear is incapacitating, causing a weakening of the limbs, a rapid heartbeat and pulse rate, dry mouth, and an inability to think rationally. Sudden fear drives the frightened being into blind panic and terror, with the expectation of injury, pain, and death, but is so debilitating and unbearable it steals the ability to move or it causes inexhaustible fleeing. It is a natural instinct of animals; in humans it is a learned involuntary reaction caused by a secretion from the adrenal glands upon stimulation by the central nervous system in response to stress, as fear. A third reaction beyond the freeze and fleeing response is anger; rage stimulating courage to fight even against overwhelming odds as in the example of Congressional Medal of Honor winners, seeing their fellow soldiers being killed, fiercely charge and subdue the enemy.

Fear in the guise of anger has been observed in military heroes who have fearlessly stormed a "machine gun nest" or overcome great odds to save fellow warriors. The same emotional anger and unstoppable violence has been witnessed in police officers reacting to a stressful situation that may not even be life threatening.

Fear will kill you. Fear will save your life. Aristotelian philosophy informs us that things have a corresponding opposite with a point in the middle, or mean, reflecting a neutral position. For example, when discussing courage, Aristotle points out that, on a linear scale, fear (or panic) is on one end and foolhardiness on the other, with courage in the middle. Foolhardiness is defined as doing something or making a decision with no redeemable features and with total disregard for one's own safety. Courage is responding in a favorable or noble way to a potentially dangerous stimulus in spite of the fear.

> That it (courage) is a mean with regard to feelings of fear … and plainly the things we fear are terrible things … people even define fear as expectation of evil. Now we fear all evils, *e.g.* disgrace, poverty, disease, friendlessness, death, but the brave man is not thought to be concerned with all; for to fear some things is even right and noble, and it is base not to fear them—*e.g.* disgrace, he who fears this is good and modest, and he who does not is shameless. He is, however, by some people called brave, by a transference of the word to a new meaning for he has in him something which is like the brave man, since the brave man also is a fearless person …. With what sort of terrible things is the brave man concerned? Surely with the greatest; for no one is more likely than he to stand his ground against what is awe-inspiring. Now death is the most terrible of all things that inspire confidence or fear in the circumstances that have been stated, and it chooses or endures things because it is noble to do so, or because it is base not to do so. But to die to escape from poverty or love or anything painful is not the mark of a brave man, but rather of a coward; for it is softness to fly from what is troublesome, and such a man endures death not because it is noble but to fly from evil.[17]

In the words of William Shakespeare's *Julius Caesar* (act two, scene 2): "A coward dies a thousand times before his death, but the valiant taste of death but once. It seems to me most strange that men should fear, seeing that death, a necessary end, will come when it will come."

In other more colloquial terms the cliché is, "A brave man dies only once, the coward dies a thousand deaths." On February 14, 2018, in a shooting at Marjory Stoneman Douglas High School in Parkland, Florida, 17 students were killed. The school had an armed presence of a deputy sheriff on site. Security cameras showed the officer waiting outside the school instead of rushing into the school to compromise the shooter. He was instantly branded a coward or "yellow" in the news media and on social sites. He was forced to resign and will always be known as a coward for failure to react according to protocol. He will live with this memory the rest of his life and die a thousand deaths.

Police risk their lives every day to protect citizens. It takes great courage to be a police officer, facing danger from criminals and gangsters, mentally ill persons, and the emotionally upset day in and day out, and yet they are expected to control fear and anger. Fear erodes confidence. The modern world has complicated policing matters; everything officers do and say is caught in pictures and sound. They must walk a fine line of being correct in their decisions, especially when in life-threatening situations when they have a thousandth of a second to decide "shoot or don't shoot." Those moments are recorded and replayed over and over, second guessed by people who were not there and are only seeing a portion of the incident out of context. One moment of an officer showing fear will define him for the rest of his entire life.

"Why did they shoot him? He was only holding a knife." The general public is unaware of repeated tests in martial arts studios and police training academies showing that a person with a knife standing within 21 feet of a police officer is a very viable threat to the life of the officer. In test after test the "attacker" with a bladed weapon reached the officer before the officer could shoot. Yet if the officer uses lethal violence, i.e. shooting the attacker, he or she will be severely criticized for "being in fear for his/her life" and taking deadly actions … "and the person was armed only with a knife."

In a dark alley or yard, maybe on a night-time street, it is extremely difficult to see and evaluate what a person has in his or her hand: Is it a gun, a knife, a toy, a phone, or a bottle of soda? The officer is in a high-stress situation and any delay in reacting or taking meditative action could easily lead to him or her being killed. Given those circumstances, and add in the adrenalin-pumping experience of driving to the scene at high speed with red and blue lights flashing, sirens wailing, and evading congested traffic, the responding officer will already be in a high alert emotion.

One citizen angrily supported the police in a comment remarking about a social media video of police using force to subdue an individual. She was angry about all the negativity she had read in the *Los Angeles Times* newspaper, seen on television and referenced in social media about police shootings and use of excessive force that she interpreted as "police brutality." She alluded to people not telling a lawyer or doctor how to do their jobs as we shouldn't tell a police officer how to do his/her job. She asked about where the line is drawn between appropriate and excessive force leading to brutality emphasizing that police lives matter.

She continued, saying that police have families and loved ones the same as everyone else. Police sacrifice their holidays, evenings, weekends, vacations, and families to help at a moment's notice. She felt she was sickened by references of police being "blasted" and second guessed on every encounter involving any type of resistance or non-compliance. She noted that unafraid police put themselves in grave danger while the news media and the public seem to want the police to wait until they are dead to react to an actual dangerous situation.

Appealingly, she said "Wake up! We need the police to keep order."

Despite what many think, police officers are subject to the same laws as everyone else and a majority of officers strive to obey them. So when the actions of a few in law enforcement are called into question, the entire profession suffers. It seems like all daily news stories involve a police shooting or use of excessive force. This has a direct effect, resulting in a lack of confidence in the police and a raising voice of "racism." Damage goes beyond the wrongfully injured or fatally wounded citizen; it hurts all those involved in the case, including law enforcement and prosecutorial staff, families of the wrongfully assaulted victim of the original alleged crime in question, and the public at large when justice is not carried out.

One citizen made the following point about police fear:

> You know what I'm sick of hearing? "Police want to go home at night too." Easy. Take a desk job! If you don't have the nerve to wait till the last minute, if you don't have the nerve to put the little bit of risk on the line that you signed up for, then go be a telemarketer, bag groceries, do guard work. But do not become a police officer.

Being fearless in the face of danger can be exhilarating. Everyone is intrigued by fear. We go to amusement parks to experience the thrill of potential danger on "space and death defying" rides; we watch movies like *Halloween, Psycho*, and *Jaws*, maybe a zombie movie. We feel the adrenalin rush and feel frightened (with an expectation of remaining safe). We don't actually expect to be thrown into a real life or death challenging event. We want to feel fear but be comfortable in the knowledge that we are enveloped in safety. That is a controllable fear and the surprises and feelings of impending doom are artificial and expected. After the experience we look forward to laughingly packing the kids up and going home. That is not the type of fear this book is about. What is discussed is the whole encompassing fear that drives a person to the limits of violent death. "The boundaries which divide Life from Death are at best shadowy and vague. Who shall say where the one ends, and where the other begins?" (Edgar Allan Poe).[18]

Fear Is a Scorpion

Everyone has fears. Fear is a scorpion residing in the brain. Fear hides in the brain until stimulated, attacking its victim with the potential to paralyze. You must give credit to a security officer working alone in a warehouse, an office building, or construction site. It takes a great amount of courage working alone to walk the dark halls, alleys, and passage ways at two o'clock in the morning, dark clouds overhead, very limited lighting, an occasional crack of thunder with rain beating against the windows. Imagine being a police officer who must enter a strange, dark building looking for a prowler. All you know is that there may or not be anyone in the building. Armed only with a flashlight and your sidearm, you walk around searching the darkness for someone…or something unknown. You are at a serious disadvantage because if there is anyone, or any creature, hiding in the darkness, it can see you and know where you are by the movement of your flashlight. This is indeed a very frightening moment. Fear can create monsters where none existed before.

Fear tricks us, a flash invading our thoughts, it knocks at the door of our mind. It says "let me in," and finding no resistance it crushes all instincts except survival and avoidance of pain. The antidote to mind-creeping fear is one's own courage, perhaps even anger; not the thinking of the fear or its origin, not the assessing of it, but the response to it. How it is responded to either erases it or makes it so overwhelming that it controls all rational thinking.

Fear and courage reside side by side in the psyche of a police officer. A very fine line separates fear and courage. Acting with bravado, an officer can mask his or her fear but on occasion the line wavers, leading to irrational actions. It can turn an otherwise stable officer into a control freak, demanding that his commands be instantly obeyed, or else it leaves the officer with a perceived loss of control, causing unreasonable reactions, making him feel insecure; the more insecure he feels, the more aggressive and unkinder he becomes, even callous, leading to unreasonable and unnecessary force. "Fear is a very important factor in intractable conflict. Emotions like fear can often cause extreme and seemingly irrational behavior in people, which can result in escalating conflict."[19]

FIGURE 2.3 A scorpion's appearance is fiercely frightening and the venom of its sting is capable of paralyzing its victim.
Photo credit: Mohamed Hassan.[20]

As the control center of the body, the brain functions as the director of traffic for all the senses. Everything we see, hear, touch, smell, taste, and perceive sends messages to the brain; it is all seeing and knowing. However, as efficient as the brain is, it has bifurcated qualities. On one hand it allows us to gain knowledge and learn, make decisions, interact with others, and a thousand other things. It also gives us a gift in that some memories fade, which allows us to forget traumatic events and unpleasant occurrences. On the other hand, the brain is not infallible, it has a weakness; it can be hijacked by fear, stifling cognitive reasoning and clear thinking.

Fear is dreadful and is capable of sucking all resistance from the brain. It can cause a person to fold up into a fetal position or stand and watch a bus bearing down on her until she becomes like a mosquito on the windshield of a speeding car. When fear controls our lives, safety and security become the main focus of all our activities.

> We all face our own struggles day to day, and must meet whatever it is. We must meet it on our own, on our personal battlefields.
>
> For a youth, that battle might be the fear of exams or being bullied in school. For an adult, it could be facing that daily commute to go to a job that is anything but inspiring. It might be not having a job at all, or failing to achieve one's dreams or goals. A new mother may fear yet another day inside with the same routine, over, and over and over again. A family too may be dealing with the looming spectre of an allergy or something worse. However small and insignificant these things may seem, they are our own battles, fears, and it's crucial that we fight on daily.
>
> No doubt that [Spartan] Leonidas and his [three hundred] men [in the battle at Thermopylae] each wrestled with some measure of fear, perhaps of loss, of not being remembered, of failing their way of life. But they overcame their fears, and though they died, they raised the bar of human achievement to heights we can only now dream of, but for which we should never cease to aim.[21]

When people have a serious medical problem they fear death. If they are in a life-threatening circumstance, they fear death. Actually there is nothing to worry about or be fearful of because death is a peaceful, quiet overtaking of the body and a relief of the pain and suffering they have been enduring. Therein is the fear of death, not the actual death, but the process of dying can be very painful and it is the feeling of hurt and pain that causes the fear of death. It is the pain of suffering prior to death they fear. When facing a serious operation such as heart surgery there should be no fear because the patient will be heavily sedated and in a deep sleep from which they may never awaken, so they will not know when they pass into an eternal sleep. Receiving a bullet or knife wound is extremely painful, yet with death comes an end to the pain. To a dying person, there should be no fear for themselves, only concern for those left behind.

Police officers, and others in high-risk occupations, may not have the luxury of dying quietly in their sleep at home, avoiding the trauma and pain of being shot. The likelihood of injury and death is a heavy burden riding daily or nightly with a police officer the moment he steps out of the police station. If he thinks of the possibility of being hurt, he will likely not react or respond as quickly as he should, thus, perhaps, endangering the lives of others. This then is also a burden he must bear.

A very serious question should be asked of all police candidates, "Would you risk your life or take a bullet for someone?" This question must be given honest consideration because the possibility does exist. Of course the answer to that question cannot be any more than hypothetical because until that very problem exists in reality, who can know how they will react? There is always a lot of bravado talk when in a safe and secure environment, but when the real scenario is played in a life-or-death situation what will the brain tell an individual to do? Will she cower behind someone else? Will she run away? Will she become frozen in blind fear? Or will she follow her training and respond the way she has practiced, and engage with the danger?

In a video of the assassination attempt against President Reagan in 1981, a police officer is seen grabbing Assistant Chief of Staff Michael Deaver, holding him in front of him and ducking down when the shooting begins. A Secret Service agent, Timothy McCarthy intentionally stepped into the line of fire to protect the president and was shot in the abdomen. The philosophy of Bruce Lee advises: "If you look within yourself and are sure you have done right, what do you have to fear or worry about?"

Let us turn again to Aristotle and his *Nicomachean Ethics* to learn more about courage and fear and how they form equilibrium on a linear scale:

> Though courage is concerned with feelings of confidence and of fear, it is not concerned with both alike, but more with the things that inspire fear; for he who is undisturbed in face of these and bears himself as he should towards these is more truly brave than the man who does so towards the things that inspire confidence.

(Aristotle meant that it is a brave man who goes forth in spite of the things causing fear, yet a lesser brave man goes forth only when he is confident he will not be injured or killed.)

> [F]or it is harder to face what is painful than to abstain from what is pleasant … death and wounds will be painful to the brave man and against his will, but he will face them because it is noble to do so or because it is base not to do so. And the more he is possessed of virtue in its entirety and the happier he is, the more he will be pained at the thought of death; for life is best worth living for such a man, and he is knowingly losing the greatest goods, and this is painful. But he is none the less brave, and perhaps all the more so, because he chooses noble deeds of war at that cost.
>
> *(Aristotle,* Nicomachean Ethics*)*

In modern armies, and perhaps in all the armies of all times, Man has voluntarily endured the struggle and pain of harsh training and faced the terribleness and horror of war because his nobleness of beliefs carried him beyond the bounds of fear of pain and death leading him to courage beyond anything real or imagined. The Congressional Medal of Honor or the "Red Badge of Courage" goes to only "the bravest of the brave and action above and beyond the call of duty." The same sentiment can be said for police officers, fire fighters, and everyday citizens who answer the call in an emergency, going beyond what is expected, gambling the risk of pain, suffering, and death to rescue another's endangered life.

In the novel *The Life of Pi*, the main character, the story teller, Pi is alone on a life boat with a Bengal tiger. In the narrative Pi expresses his thoughts of fear:

> I must say a word about fear. It is life's only true opponent. Only fear can defeat life. It is a clever, treacherous adversary, how well I know. It has no decency, respects no law or convention, shows no mercy. It goes for your weakest spot, which it finds with unerring ease. It begins in your mind, always. One moment you are feeling calm, self-possessed, happy. Then fear, disguised in the garb of mild-mannered doubt, slips into your mind like a spy. Doubt meets disbelief and disbelief tries to push it out. But disbelief is a poorly armed foot soldier. Doubt does away with it with little trouble. You become anxious. Reason comes to do battle for you. You are reassured. Reason is fully equipped with the latest weapons of technology. But to your amazement, despite superior tactics and a number of undeniable victories, reason is laid low. You feel yourself weakening, wavering. Your anxiety becomes dread ... For fear, real fear, such as shakes you to your foundation, such as you feel when you are brought face to face with your mortal end nestles in your memory like a gangrene; it seeks to rot everything, even the words with which to speak of it. You must fight hard to express it. So you must fight hard to shine the light of words upon it. Because if you don't, if your fear becomes a wordless darkness that you avoid, perhaps even manage to forget, you open yourself to further attacks of fear because you never truly fought the enemy who defeated you.[22]

Summary

Fear has many faces and comes in several packages. One person's fear may be another person's thrill ride. If suddenly confronted by danger, such as a stalking mountain lion, one person may turn and try to flee, but is quickly overtaken by the lion and killed, or he may be frozen in place waiting for the lion to pounce. Another person in the same situation may raise her arms, rising to her highest height and charge the lion while growling and screaming, causing the hungry lion to turn and run. Reactions to fear may save a life or may result in unfavorable consequences.

Fear is a scorpion of the brain having the power to paralyze; freezing the body in place or increasing adrenalin, stimulating the fight reflex and aggression. Fear affects people in various ways. It is a strong reactive emotion, when people feel their life is in danger and they have no control over their security. Courage is at the midpoint between foolhardiness (careless and reckless) and cowardice. It is an unseen line and may be invisible until that telling moment when a life is threatened or placed in a dangerous circumstance.

Courage is looking at the face of death and spitting in its eye.

Review Questions

1. Explain why fear is the ultimate unknown weapon.
2. How can fear cause a well-trained officer to over-react with force and violence in a potentially dangerous situation?
3. "A coward dies a thousand deaths." Explain.
4. What is the meaning of the Charles Dickens quote?
5. Do you believe fear is a cause for a police officer to over-react?
6. Summarize the people, circumstances, and conditions affected by violence and those who would be its perpetrators and victims.

Notes

1 www.bing.com/search?q=synonyms+of+fear&qs=n&form=QBLH&sp=-1&pq=synonyms+of+fear&sc=8-12&sk=&cvid=2F5222BC024F4C84B44BCD1B24F14578. Retrieved January 8, 2019.
2 *Webster's Dictionary* 1942. New York: Publishers Guild.
3 www.listofhumanemotions.com/fear. Retrieved July 9, 2019.
4 https://answers.yahoo.com/question/index?qid=20070926175441AAy81fv. Retrieved October 21, 2019.
5 www.goodreads.com/quotes/tag/fear-of-unknown. Retrieved December 16, 2018.
6 Steimer, Thierry; "The Biology of Fear- and Anxiety-Related Behaviors". Dialogues in Clinical Neuroscience, 2002 Sep, 4(3): 231–249. www.ncbi.nlm.nih.gov/pmc/articles/PMC3181681/. Retrieved September 22, 2019.
7 McFarland, D. *The Oxford Companion to Animal Behaviour*. Oxford, UK: Oxford University Press; 1987 as referenced in "The Biology of Fear and Anxiety" by Thierry Steimer, PhD, *Dialogues in Clinical Neuroscience*. 2002 September; 4(3): 231–249.
8 www.listofhumanemotions.com/fear. Retrieved July 9, 2019.
9 Quote from David Gemmell (2005). *Troy: Lord of the Silver Bow*. London: Corgi.
10 www.psychologistanywhereanytime.com/phobias_psychologist_and_psychologists/psychologist_fears_and_phobias.htm. Retrieved July 9, 2019.
11 http://psycheducation.org/brain-tours/fear/. Retrieved September 7, 2018.
12 Kabat-Zinn Jon (2013). *Full Catastrophe Living* (Revised Edn): *Using the Wisdom of Your Body and Mind to Face Stress, Pain, and Illness*, p. 521, Bantam. www.azquotes.com/quote/535475. Retrieved March 5, 2019.
13 www.facebook.com/theDanRather/posts/10158192562480716. Retrieved February 12 2017.
14 www.online-literature.com/dickens/2787/. Retrieved July 27, 2014.
15 Beginning in 1940, a man named George Metesky hid 33 pipe bombs in public spaces in New York City. Twenty-two of those bombs exploded, injuring 15 people. Until he was captured in 1957, Metesky was known to the press, the police, and an increasingly anxious populace as the "Mad Bomber." www.npr.org/2011/05/13/136287845/a-16-year-hunt-for-new-yorks-mad-bomber. Retrieved March 22, 2018.
16 Hanna, Jason, Faith Karimi, Jason Morris, and Steve Almasy; CNN, August 31, 2018 *Police: Austin Bomber Left 25-Minute Confession Video on Phone*; www.click2houston.com/news/national/austin-bombers-motive-is-to-cause-mayhem-and-death. Retrieved October 20, 2018.
17 Aristotle, *Nicomachean Ethics;* Appearing in *The World's Great Thinkers; Man and Men; The Social Philosophers*; Random House, New York; 1947.
18 www.brainyquote.com/quotes/authors/e/edgar_allan_poe.html. Retrieved July 27, 2015
19 Barker, Phil, www.beyondintractability.org/essay/fear/. Retrieved March 8 2017.
20 https://plixs.com/photo/5629/scorpion-silhouette.
21 *Facing Fear with the 300 Spartans*; Eagles and Dragons Publishing; 2015; http://eaglesanddragonspublishing.com/facing-fear-with-the-300-spartans/. Retrieved January 1 2017.
22 Yann Martel, *The Life of Pi*; Houghton Miffin, Harcourt Publishers, Orlando, FL. 2001, pp. 161–162.

3

ANTECEDENTS OF FEAR

Fear and the Brain

The amygdala is an almond-shaped set of neurons located deep within the brain's medial temporal lobe. It plays a key role in processing emotions, and is linked to both fear and pleasure in humans and other animals. The amygdala is also correlated with aggressive behavior over a number of species. Conditions such as anxiety, autism, PTSD, depression, and phobias are thought to be associated with abnormal functioning of the amygdala.[1]

During normal perception, our sense organs transmit stimuli to part of the brain called the thalamus en route to the sensory cortex, where perceptions and thoughts are generated. To have a visual perception, for instance, "information has to be transmitted from the eye, from the retina, through the optic nerve, into the visual thalamus, and from the visual thalamus to the visual cortex, where the processing continues and you can ultimately have the perception," says Joseph LeDoux. "The visual cortex connects directly with the amygdala, and so that one route by which the information can get in: retina, thalamus, cortex, amygdala."

But LeDoux, one of the foremost researchers of emotion and fear in the brain, discovered early in his research that fear, despite being reliant on sensory stimuli to trigger the amygdala, was not processed via the normal sensory pathways, or at least not entirely. "What we found was that if the cortical pathway was blocked completely. Rats could still form a memory about a sound." Instead of traveling from ear to thalamus to cortex to amygdala, the loud sound could bypass the cortex altogether, going directly from thalamus to amygdala. "This was really important because we generally think that the cortex is required for any kind of conscious experience," LeDoux says. In other words, a frightening stimulus can trigger emotions and fears unconsciously—without our even realizing it.[2]

Most fears are learned, except the fear of loud noises.

Even newborns are frightened by loud noises. The response to an unexpected loud noise, known as the startle reflex, is extremely fast: Auditory information goes first to the

> cochlear nucleus in the medulla and from there directly to an area in the pons that commands the tensing of the muscles .... Information reaches the pons within 3 to 8 milliseconds after a loud noise, and the full startle reflex occurs within two-tenths of a second (Yeomans & Frankland, 1996).[3]

Although the fear of loud noises is not learned, but innate, the startled person's current mood or past experiences can affect the reaction. A more powerful startle reflex is triggered in someone who is already feeling tense, as in the instance of a police officer approaching a potentially dangerous situation or an unknown circumstance. Startle being a precursor to fear.

Irrational fears are wrapped in imagination and unreasonable foundations. They are of phantom thoughts with no basis for the fear requiring avoidance or a change of normal habit. In short, irrational fears are more known and diagnosed as phobias.

Phobias

"A phobia is an overwhelming and unreasonable fear of an object or situation that poses little real danger but provokes anxiety and avoidance ... a phobia is long lasting, causes intense physical and psychological reactions, and can affect your ability to function normally at work or in social settings."[4] The key words in this definition are *overwhelming* and *unreasonable*. With a phobia there is no instant threat of danger. Thoughts invoked by some irrational association with a circumstance or situation can create fear for a phobic that is as real as any fear experienced by a non-phobic.

Symptoms of a Phobia

Symptoms of a phobic reaction are very similar to the reaction caused by actual fear, with biological changes to the body and mind. A phobic reaction is a feeling of uncontrollable panic, terror, or dread when exposed to the source of fear.

- The feeling that you must do everything possible to avoid what you fear.
- The inability to function normally because of your anxiety.
- Physical as well as psychological reactions, including sweating, rapid heartbeat, difficulty breathing, a feeling of panic and intense anxiety.
- Often, the knowledge that your fears are unreasonable or exaggerated but feeling powerless to control them.
- In some cases, anxiety just thinking about what you fear.
- In children, possibly tantrums, clinging or crying.[5]

Obsessive compulsion behavior is often related to phobias, for example the constant washing of hands, or using a napkin or glove to turn a door knob because of the fear of contamination. Contamination fear phobias, variously referred to as misophobia, mysophobia, or molismophobia, are most commonly referred to as anxiety disorders. These are characterized by an "overreaction to the slightest uncleanliness or an irrational fear of dirt, contamination, or defilement"[6] and fear of "being contaminated with dirt or germs."[7] The fear of contamination has become so pervasive in society that it is not uncommon to see all types of workers wearing latex gloves in performance of their duties.

There are many types of phobias listed in medical journals, some of which are:

- the fear of being touched (Haphephobia);
- the fear of death (Thanatophobia);
- the fear of failure (Atychiphobia);
- the fear of fear (Phobophobia).[8]

Phobias are long lasting and unreasonable but can be successfully treated with cognitive behavioral therapy and medication. All of these phobias may be an unconscious weakness in all but the most confident of police officers. The last three phobias, fear of death, fear of failure, and the fear of fear in a police officer may be motivating factors causing him or her to overcompensate in many, even normal, circumstances.

Paranoia

Generally speaking, an officer has two types of paranoia. One ten-year veteran officer stated: "It really has come down to two things; fear of failure and failure of not being in control." These have nothing to do with mental disorder, fear of bodily injury, or the feelings of being the target of someone else's fear of police. Perhaps, to be more accurate, they manifest as a constant state of anxiety.

"They are out to get me!" There is a comedic cliché saying, "Just because you are paranoid doesn't mean they aren't out to get you." "Paranoia is an unfounded or exaggerated distrust of others, sometimes reaching delusional proportions. Paranoid individuals constantly suspect the motives of those around them, and believe certain individuals, or people in general, are out to 'get them.'"[9]

Paranoia is characterized by delusions of persecution, unwarranted jealousy, or exaggerated self-importance. It may be an aspect of chronic personality disorder, of drug abuse, or of a serious condition such as schizophrenia in which the person loses touch with reality. Paranoid individuals find it very difficult to get along with other people because of their own hostility and aggressiveness. They feel that all opposing people are spying on them.

A sufferer of paranoia will generally behave very aggressively with other people. This in turn elicits aggressive behavior from the opposing person, which again reinforces the paranoia of the individual who has a serious, usually ungrounded suspicion and mistrust of people or their actions without evidence or justification. A paranoid individual, being suspicious of the motives of the persons around her, expresses unreasonable fear of what can be described as irrational reaction to an innocuous person, group, or circumstance. A certain degree of paranoia is a part of the "police personality." Not to the degree of being psychotic but having a suspicious or curious nature beyond that of the "average" person.

Paranoid fear can become overwhelming and unshakeable, leading to hallucinations and bewildered thinking, often opening the door to further deterioration and schizophrenia.

Paranoid schizophrenia is a fantasy experience of what is known as persecutory delusions. A paranoid schizophrenic, reacting to the delusion and fear he is suffering may very likely strike out with intent to hurt or kill the object of his fantasy. They can be dangerous … very dangerous!

Depression

Depression is not normally considered a precursor to fear or a threat, but under certain conditions a police officer must be wary of someone who is bi-polar (manic-depressive). A depressed person might appear weepy and subdued, very docile. Anger is a lion when it comes to transiting from the depressed stage to mania. In the mania state, anger is a fierce motivator to aggressively strike out verbally or physically. An unsuspecting officer might be lulled into believing the depressed person is not of threat and will be taken by surprise when the person becomes manically animated and strikes out in anger or fear.

Worry, Anxiety, Fear, Terror, Panic

There is a linear continuum from simple worry to blind overwhelming panic. "A person often tries to categorize what they experience, but essentially they're just different levels of the same state. A sort of 'anxiety scale', with worry and mild concern at one end of the spectrum, progressing through anxiety, to fear and blind panic at the other."

The complete fear emotional experience chain follows a linked progression of anxiety, dread, startle, fear, terror, and panic. Anxiety is a "crashing" of the nervous system, potentially leading to physical distress, nausea, or feelings of panic in certain circumstances. It is often alluded to as "being nervous." Dread is a feeling of worry or anticipation of something that might happen that will cause pain or unpleasantness, such as going to the dentist for a root canal or waiting to hear from a doctor the results of a biopsy, or perhaps in anticipation of a court appearance. A startle is a sudden and totally unexpected frightening experience presenting an instantaneous involuntary reaction to an emotional sensation of immediate danger. It causes a momentary spurt of adrenalin that dissolves as soon as the brain analyzes the circumstances and sends the message that there is nothing to fear. If the startle is not quickly resolved, it can evolve into fear with full physiological and psychological consequences. The sudden flood of adrenalin, noradrenalin, and dozens of other hormones causes changes in the body, which include:

- Heart rate and blood pressure increase.
- Pupils dilate to take in as much light as possible.
- Veins in skin constrict to send more blood to major muscle groups (responsible for the "chill" sometimes associated with fear—less blood in the skin to keep it warm).
- Blood-glucose level increases.
- Muscles tense up, energized by adrenaline and glucose (responsible for goose bumps—when tiny muscles attached to each hair on surface of skin tense up, the hairs are forced upright, pulling skin with them).
- Smooth muscle relaxes in order to allow more oxygen into the lungs.
- Nonessential systems (like digestion and immune system) shut down to allow more energy for emergency functions.
- Trouble focusing on small tasks (the brain is directed to focus only on the big picture in order to determine where threat is coming from).[10]

FIGURE 3.1 Anxiety/panic sequence[11]

If the continuum utilized a color code, Worry would be purple, Anxiety yellow, Fear red and Panic black. Worry would be purple because worrying leads to feelings of helplessness, negative thinking, and perhaps depression ("the blues"). Anxiety becomes yellow because it manifests jumpiness and feeling on edge, heightened senses, looking out for danger, muscles tensed, just in case physical action is required. It becomes a heightened sense of worry of what is in the future. It can be reckoned as living in the future as nothing has occurred but the feeling builds because something may occur. It is a way to prepare the body and mind for what may or will happen. Fear is an advancing orange to a full red involving thoughts of survival and restriction of the reasonable thought process. Panic should be considered as the color black because all reasoning is lost in a blind rush, with tunnel vision to escape the deadly threat as quickly as possible, seeking means of survival. It completely takes control of all reasoning and senses, to the point of having no concern for others in a common circumstance.

> Fear is often confused with anxiety, which is an emotion that is often exaggerated and experienced even when the source of danger is not present or tangible. While fear is connected to anxiety and other emotional conditions, like paranoia and panic, it is a separate emotion on its own.[12]
>
> Fear and anxiety are closely related. Fear is associated with a strong tendency to escape from some immediate danger or threat. Anxiety is a general sense that something dangerous might occur but it is unknown what, when, where, or how. There might even be an attempt to determine the cause of the anxiety.[13]
>
> Fear and anxiety often cause very similar symptoms, such as muscle tension, increased heart rate and shortness of breath brought about by the body's flight-or-fight instinct. Fear is referred to as an emotional response to a situation in which an individual feels threatened. The cause of the threat is realistic in nature. Oftentimes, fear of a particular situation or event is caused by a traumatic event experienced earlier in life. The effects of this traumatic event are carried by the individual throughout his or her life to such an extent that when the individual finds himself or herself in a similar situation, he or she begins to exhibit the symptoms mentioned above.[14]

Terror is bone-chilling, heart-stopping, and thumping fear. It is a blind, uncontrollable reaction to a life-threatening occurrence over which the individual has no control and death is perceived as imminent and inescapable. Imagine a passenger on an airliner at 38,000 feet when suddenly the oxygen breathing masks drop down and the plane goes into a steep, straight-down vertical dive and an engine is on fire. Imagine being on one of the planes that crashed into the World Trade Center or being a pilot on the plane and seeing the ground or building quickly approaching and there is nothing that can be done to stop or change course.

Panic, of course, is that instant of blind terror when adrenalin is pumped so powerfully into the system that the brain stops its cognitive functioning, vision closes to a pinpoint or shuts down, and the affected person wildly screams and attempts to flee; not seeing, knowing, or caring where she is running so far as it removes her from the life-threatening situation. Often panic will lead the person to run into the source of her fright or induce panic in others. Panic is closely related to terror. It is the losing of all emotional and physiological control except for the overpowering urge to escape and run away in a blind direction with not even a thought of anything except getting as far away from the threat as fast as possible. Tunnel vision will result in a darkness, rendering the panicked individual to be susceptible to running into even greater danger.

Stephen King, noted for his blood-curdling horror stories, is quoted as saying:

> The 3 types of terror: The Gross-out: the sight of a severed head tumbling down a flight of stairs, it's when the lights go out and something green and slimy splatters against your arm. The Horror: the unnatural, spiders the size of bears, the dead waking up and walking around, it's when the lights go out and something with claws grabs you by the arm. And the last and worse one: Terror, when you come home and notice everything you own had been taken away and replaced by an exact substitute. It's when the lights go out and you feel something behind you, you hear it, you feel its breath against your ear, but when you turn around, there's nothing there.[15]

The terms are used interchangeably. The basic difference is:

> **Worry and anxiety**—A set of responses to an unknown, imprecise or ill defined threat; often anticipatory in nature and created by the imagination. It's more associated with the need to be prepared. Worry leads to feeling anxious.
>
> **Fear and panic**—A set of responses to a known, precise, well defined threat, which can be real or vividly imagined. It's mainly about avoidance and escape. In its extreme form, fear becomes panic.[16]

A sudden uncontrollable fear or anxiety, often causing wildly unthinking behaviour, panic, defined by psychologydictionary.org, "is an abrupt, unmanageable fear response which might consist of horror, confusion, and absurd behavior, caused by an assumed threat."[17]

The Anatomy of Anxiety

> Anxiety is considered to be a psychological disorder where the individual experiences symptoms similar to those experienced by those who face fear-inducing situations or circumstances. The difference between anxiety and fear is that, unlike fear, symptoms leading to anxiety occur even though there is no apparent risk or cause for physical harm. More often than not, the reason the individual feels anxious cannot be pinpointed. This is in stark contrast to fear, where the individual can readily determine the root cause of their fear. Those who suffer from anxiety find themselves helpless and unable to cope with their symptoms to such an extent that it begins to interfere with their daily activities and interactions with other people. Anxiety is often one of the primary causes of other psychological disorders, such as depression and personality disorders.[18]
>
> Anxiety [a synonym for worry or nervousness] is a reaction to stress that has both psychological and physical features. The feeling is thought to arise in the amygdala, a brain region that governs many intense emotional responses. As neurotransmitters carry the impulse to the sympathetic nervous system, heart and breathing rates increase, muscles tense, and blood flow is diverted from the abdominal organs to the brain. In the short term, anxiety prepares us to confront a crisis by putting the body on alert. But its physical effects can be counterproductive, causing light-headedness, nausea, diarrhea, and frequent urination. And when it persists, anxiety can take a toll on our mental and physical health.[19]

Excessive stress and worry related to work, especially in a hazardous occupation such as policing, firefighting, or coal mining, plays on the unconscious sleeping mind, resulting in a form of disturbing sleep and real lifelike dreams, termed "anxiety dreams." This type of dream reveals aspects of fear such as being in a gunfight and being unable to pull the trigger, falling through the roof of a burning building, or experiencing a mine caving in and being trapped in deep darkness, surrounded by nothing but rocks and with no way to escape.

In response to the question of fear while working in a dangerous environ one female private investigator, former police officer, and certified executive protection specialist (Monica Duperon) stated:

> I wouldn't say it's fear per se. Being overwhelmed is not being in fear. I am overwhelmed and have a lot going on but I am not afraid of it. Fear? I have felt fear, yet at the same time have been fearless. To be fearless is not that you are unafraid but rather that you acknowledge risk and being fearless is that you face it in spite of the fear of risk. I can tell you when I was working in Africa as an executive protection agent I was in areas where I felt fear, however due to my faith (training and experience) I also felt a sense of protection. It was as if I could see and feel the dangers around me yet I felt as if in a bubble which allowed me to feel fearless. I can say that true fear in the full sense of the word has been felt in the nightmares triggered during a peaceful night tormented by the deep rooted anxieties of reality which one has chosen not to address while awake.[20]

You are walking with your partner after a late-night dinner at a very nice restaurant when a large, tattooed, unshaven "biker type, weight-lifting skinhead" wearing a denim jacket with cut-off sleeves, steps in front of you and displays a knife in his hand and demands your watches, rings, and wallets. His appearance is so sudden you are startled and you stand there, waiting for your mind to clear with this sudden threat to you and your partner. Your confidence is momentarily lost and you can't move or say anything except to maybe stutter. You are experiencing what is called an "adrenalin dump."

Now you are a police officer answering a domestic violence call. It's ten o'clock at night and the street is dark, save for a street light halfway up the block. As you park and exit your car, you see a man carrying a "shoulder gun" run from the house toward an alley behind the house. The yard is dark, barely lit by ambient lighting in the alley. You can see a tall fence at the back of the yard but the man is out of sight. Suddenly you hear the unmistakable sound of a round being pumped into the chamber of the man's gun. You turn to face him about seven yards away. Adrenalin, which is already running high, pushes to new levels, fear controls your thoughts. Shoot or don't shoot?

So what is adrenalin and what does it do? Adrenaline is a hormone and neurotransmitter. Its purpose is to physically prepare humans for fight or flight. When we are physically attacked, or even mentally stressed by someone being aggressive, this hormone is dumped into our bloodstream by the adrenal glands.

Immediately the heartbeat starts to climb. Blood rushes to the larger muscles of the body and away from the extremities and brain. These muscles become extremely strong (we have all heard the stories of a mother lifting an automobile off her child, the obvious effects of an adrenal dump). You may feel weak. Your body may shake. Your throat may become very dry. It may be hard to speak normally. You may even feel sick to your stomach. You will most likely lose most of your fine motor coordination. This is partially due to the fact that your blood flow to the fingers and hands may be restricted. At first your sight may be extremely sharp as your pupils dilate. But quickly it pinpoints in on one area, like looking through a small tube or funnel.

The higher your heartbeat becomes, the less likely it will be that your cognitive thinking will be clear. You may start to lose the capacity to hear. The ability to remember things is hampered. Time seems to be suspended. If your heartbeat goes too high, you may completely freeze like a deer in the headlights.[21]

Shoot or don't shoot? You want to command him to "Drop the gun!" but your voice comes in a raspy croak. Trying to inhale a deep breath, you loudly say again, "Drop the gun!" You note he is holding the gun upward at an angle and realize that to shoot you he will have to move the gun into a firing position. In that thousandth of a second you will discharge your weapon. Fear runs through your body. Time has stopped. To you, it feels like several minutes have passed.[22] Your breathing comes in short bursts. Your pistol with a trigger pull of two and a half pounds is in your hand; your trigger finger begins to tense. Your eyes, narrowed to tunnel vision, see only the man's chest aligned with your back and front sights. Any slight movement he makes will be your cue to continue pressure on the trigger of your gun. The next breath he takes likely will be his last. It is your decision, does he live or die? Time has not elapsed beyond thousandths of a second, not the several minutes you imagine. Shoot or don't shoot? He drops the weapon; your physiological being starts restoring to normal. You are still reacting with adrenalin at a high state of awareness as you move in to subdue him. Relieved neither he nor you decided to shoot. Fear of the life-threatening circumstances has passed.

You realize if you had pulled the trigger, he would have died and the shooting would have been ruled justifiable because you were "in fear for your life." At the same time, you also know that "all guns are always loaded" and in those tension-filled seconds, you could have been killed. Police are faced with a similar dilemma whenever they confront a person holding a gun.

In a real-life circumstance reflecting this scenario, one night, when the author of this book was a police officer, a dispatched call of "family disturbance at (a specific address) with gun involved" was received by my partner and me. As we arrived at the address, a middle-aged black man carrying a shotgun was seen running from the house and into the rear yard in an attempt to get away. My partner ran around the back of the house to an alley behind the house in an attempt to stop the man. I (perhaps foolishly) followed the man into the dark back yard. Holding my flashlight high and out in my left hand, and my .357 combat magnum revolver in my right, I saw the man stopped with his back to the fence and facing me.

Adrenalin flowed and my vision closed to tunnel vision so all I could see was the man holding the gun and the sights of my gun in perfect alignment with his chest. I took a deep breath and exhaled slowly. I repeated in my head, "Relax and don't jerk the trigger." The rest of my senses were totally focused on the moment and I was only aware of the man and my sights. Twice I ordered him to "drop the gun"; both times he didn't. I had to make an instant decision, "shoot or don't shoot." The reasoning in my mind was that for him to shoot me, he would first have to bring the gun into firing position from how he was holding it. I rationalized that in the first flicker of movement with the gun I would be able to squeeze off my round with my "hair trigger" revolver. On the third command, the man dropped the shotgun to the ground and sat down and began crying.

I picked up the shotgun and called my partner to come into the yard. The story the man told us made me feel fortunate that I had not chosen to shoot, though it would have been a justifiable shooting because I could swear "I was afraid for my life and felt immediate endangerment." The man said he and his wife were going through a difficult divorce and he was just trying to take the shotgun his grandfather had given him when he was a boy, and his wife wasn't giving him the gun so he went to retrieve it. The shotgun was unloaded! No shots were fired, no one died, but two people very easily could have, and no charges were filed. Thus, a happy ending to something that could have had a tragic result coming from a misunderstanding, fear, and lack of patience.

A valuable lesson was learned. Merely because a weapon is seen is not cause to believe a life is threatened or in immediate danger. A complete instantaneous threat assessment must be made in those thousandths of a fraction of a second!

Summary

Fear activates various elements of the brain, most notably the almond-shaped amygdala responsible for emotional responses. Irrational fears are wrapped in imagination and unreasonable foundations. A linear continuum of worry, anxiety, fear, terror, and panic illustrates a psychological progression from mild worry and dread to full-fleshed terror and panic.

In a "shoot or don't shoot" situation there is very little time to make a full assessment of the degree of danger to a police officer or a bystander. In thousandths of a second an officer must make a life altering decision that will affect many people. The officer will experience many emotions in those fractions of a second that can have lethal consequences.

Review Questions

1. Describe the anatomy of anxiety.
2. What effect can phobias have on the actions of a police officer?
3. How does the adrenalin "dump" impact the person experiencing it?
4. What are anxiety dreams? Give original examples.
5. Name the color code of the fear–panic continuum.
6. Discuss the variation between emotions and instinct.

Notes

1 www.quora.com/Which-parts-of-the-brain-control-which-human-behaviours. Retrieved September 18, 2018.
2 Miller, Max; *How the Brain Fears.* (September 22, 2010). https://bigthink.com/going-mental/how-the-brain-fears. Retrieved December 19 2018.
3 Kalat, James W.; *Biological Psychology, 9th Edn;* Thompson/Wadsworth Publishers, 2007 p. 366.
4 www.mayoclinic.org/diseases-conditions/phobias/basics/definition/CON-20023478. Retrieved August 3, 2015.
5 Ibid.
6 http://medical-dictionary.thefreedictionary.com/Germaphobia. Retrieved January 21, 2017.
7 www.common-phobias.com/myso/phobia.htm. Retrieved September 22, 2019.
8 Phobia List—The Ultimate List of Phobias and Fears; Fearof.net; www.fearof.net/. Retrieved September 10, 2016.
9 http://medical-dictionary.thefreedictionary.com/paranoia. Retrieved August 3, 2015.
10 http://science.howstuffworks.com/life/inside-the-mind/emotions/fear2.htm. Retrieved January 21, 2017.
11 *Worry, Anxiety and Fear;* www.psychology-solution.com/anxiety/worry-anxiety-fear-panic. Retrieved September 19, 2018. The term "terror" added by the author. The original anxiety/panic sequence was void of the term.
12 www.listofhumanemotions.com/fear. Retrieved July 16, 2015.
13 Kalat, p. 366.
14 *Difference Between Fear and Anxiety: Fear vs Anxiety.* www.differencebetween.net/language/difference-between-fear-and-anxiety/. Retrieved September 22, 2019.

15 www.goodreads.com/quotes/84666-the-3-types-of-terror-the-gross-out-the-sight-of. Retrieved July 27, 2015.
16 *Worry, Anxiety and Fear*, www.psychology-solution.com/anxiety/worry-anxiety-fear-panic. Retrieved September 19, 2018.
17 https://psychologydictionary.org/panic/. Retrieved June 21, 2019.
18 *Difference Between Fear and Anxiety: Fear vs Anxiety*. www.differencebetween.net/language/differ ence-between-fear-and-anxiety. Retrieved September 8, 2015.
19 www.health.harvard.edu/staying-healthy/anxiety_and_physical_illness. Retrieved March 19, 2017.
20 Monica Duperon Rodriguez—Progressive Investigations Research and Consulting Corp.—Tampa/ St. Petersburg, Florida Area, responding to a survey question.
21 Deer, as many other animals, have a survival instinct that, when endangered, they stand completely still to blend into their surroundings.
22 "One of the seats of emotion and memory in the brain is the amygdala [David Eagleman] explained. When something threatens your life, this area seems to kick into overdrive, recording every last detail of the experience. The more detailed the memory, the longer the moment seems to last. "Burkhard Bilger, "The Possibilian." *New Yorker Magazine*, April 18, 2011. www.newyorker. com/magazine/2011/04/25/the-possibilian. Retrieved June 21, 2019.

4

HANDLING FEAR

How can people shrug off fear's paralyzing effects? Dale Carnegie, an American writer and lecturer and the developer of famous courses in self-improvement, salesmanship, corporate training, public speaking, and interpersonal skills, spoke of fear as, "You can conquer almost any fear if you will only make up your mind to do so. For remember, fear doesn't exist anywhere except in your mind."[1] If the mind doesn't register a perception of danger, the eyes won't see it. In other words, the eyes can't see what the brain doesn't know. A looming specter of injury or death may be upon a person, but until she acknowledges it, she is not fearful. Seeing something and not responding is not the same as seeing something, recognizing and responding to it. The secret is recognizing what a person sees. He may see (or think he sees) danger in an innocuous circumstance, but as soon as his mind recognizes there is no threat, fear passes into relief.

With fear originating in the brain, that, obviously, is the place to begin controlling fear. Acknowledging fear, especially from the unknown, is living in the future because fear based on what may happen, and nothing has yet happened to warrant fear of injury or death, is merely unnecessary worry or anxiety. Controlling anxiety and worry caused by fear of what may happen is to look at it and ask: "What am I afraid of?"; "Is fear justified?"; "Is the danger real or am I imagining it?"; "Am I letting my imagination grow unreasonable fear?"; "Am I stressing too much about nothing or over something that has not occurred?"; "Am I over analyzing the thing causing me fear?"; "What is the worst thing that can happen?"; "Why am I frightened of this?"; "Are other people sharing my fear?"; "Am I being reasonable and am I acknowledging fear of something that has no power to hurt me?"; "Am I vulnerable to what I am afraid of?" When you are afraid of anything, you are acknowledging its power to hurt you. In your fearlessness, you acknowledge your invulnerability.

Defeating Anxiety and Fear

Suggested steps for defeating anxiety and fear:

- The next time you feel your heart pounding from feeling fear: Relax—close your eyes, focus on breathing, focus on making your breath even, take deep inhales, slow exhales, control your breathing, let the fear wash over you, don't resist.

- Breathe—Regulate breathing to lower the heart rate and blood pressure. Take long, slow, deep breaths; in through the nose (deep breathing originates in the abdomen, not in the upper chest) and out through the mouth. Fast, shallow breaths exacerbate the problem by elevating the physiological changes in the brain and body. On the intake breath, pretend you are holding a large heavy ball with both hands and arms fully extended. Then as you breathe in, slowly raise the heavy ball to chin level (count to four), Hold there and count to four again. Now flip your hands over and push the ball toward the ground (counting to four) while slowly, but forcefully, exhaling and expelling all the air in the lungs. Repeat at least four times with the eyes closed. Caution! Don't attempt this type of breathing if danger is imminent. However, taking slow, deep breaths will help relax the body and prepare it for action. (This is also good advice if you are trying to relax and go to sleep.)

In a fight or flight context, breathing should be shallow intake and complete forceful exhale. Breathing not only relaxes the body but it also helps focus the mind and helps control the adrenalin rush. Someone being hit would expel the breath forcefully to absorb the power of the punch.

- If the issues addressed in the questions stated in the previous paragraph are resolved favorably and deep understanding is achieved, fear will exist as a phantom emotion.
- Engage the anger emotion and believe the fear will not hurt you. Anger overrides most other emotions and will elevate the frightened person with a burst of energy to fight what is frightening him or her.
- Awareness—Being aware of what there is to be frightened about helps in acting normal, which, in turn, lowers anxiety.
- Visualization and mental rehearsal—Picture yourself in the frightening situation and visualize how you will act (react) when it actually happens.
- Knowledge—The more a person knows about the fear causing stress and anxiety, the easier it is to negate it.
- Self-control—Relax, mentally walk yourself through the possible scenario to control the arousal instincts. This can be accomplished by mentally talking to yourself and attempting to describe to yourself all possibilities. While the mind is occupied talking to itself about the fear, the rest of the body resists the physiological change and prepares the adrenalin necessary to overcome the fear.
- Slapping—When the adrenalin rush begins, slap your chest (solar plexus area) three or four times while continuing the short breathing and forceful exhaling technique.

> Risk Taking. I tend to "play it safe" in almost everything I do. I'm not sure if it's because of a fear of failure, a fear of success or even a fear of the unknown. Whatever my reason, I know it is because of fear, that I hold back. By facing my fears then I can become bolder in my approach to the unknown. The only way to do something you've never done before is to just do it! No amount of talking about it will get it done. Even if my boldness is fake, at least I'll have some place to get started!
>
> *National University Student*

How to Cope With Extreme Fear

1. Immediate effects—If you feel the initial effects, try to remember the way in which you handled the incident, and how you coped with the feelings, and then try to believe that you can count on these abilities to use in the future. By thinking of how well you

coped with this event, you will become aware that you will be able to cope with any future stressful event.

2. Short-term effects—Avoid being critical, and do not emotionally punish yourself. Notice how often you experience terrifying emotions or flashbacks. You may want to learn new strategies and techniques to soothe yourself, like change whom you spend time with or find ways to minimize overwhelming emotions. Try spending time with family or friends, writing in a journal, or learning a new physical activity. These intense emotions should gradually diminish within a month or two.
3. Long-term effects—If the emotional distress does not fade, if the intensity increases, or if the distress interferes with the daily routine, professional assistance is recommended. Be kind to yourself and accept the fact that you are going through a difficult time at the moment, and decide that you are going to find ways to manage this problem.

In conclusion, those of you who have lived through a trauma can help yourselves by being with caring friends and family. Take care of yourself physically, by eating well, exercising, and getting adequate sleep. To overcome trauma it may also help to write thoughts down in a daily journal (writing is therapy as it stimulates the serotonin and endorphins in the brain) or attempt to learn a relaxation exercise just to ease the mind of the tension.[2]

Fear is the result of the presence or perception of a menacing danger presenting the possibility of serious bodily injury or life threatening peril. "The human brain may respond differently to threats based on proximity, trajectory, and expectations," according to a study by Dean Mobbs and colleagues.[3]

The Mobbs study involved participants whose reactions were electronically measured and recorded as a tarantula moved closer to or further away from them. Obviously as the threat lessened when the giant spider moved away, participants were shown to exhibit less fear. If the threatening tarantula turned and moved toward them, their heart rates, respiration, and blood pressure elevated accordingly.

Fear is also the child of perception. If a person walks on a narrow board as it lies on the ground, she can fearlessly maintain her balance and perhaps even dance across it. If the board is raised higher and higher, the person's fear of falling grows until she reaches a height where she is no longer able to take one step onto the board.

In a humorous instance a person was receiving a CAT scan, which involves being placed on a sliding bed platform and moved into a very narrow chamber. There is no room even to raise or move a hand. The chamber is within inches of the patient's face and body. The technician offered the patient an option of a blindfold while being scanned. The blindfold was refused as no fear of being in tight places had ever been experienced and the patient wanted to display his boldness.

With the patient's eyes open, a fear or emotion known as "claustrophobia," which he had never previously experienced, enveloped him and he yelled to the practitioner, "Get me out of here! I can't stand it." He associated his inability to move to being encased in a casket in a grave. After being removed and a blindfold placed over his eyes, the scan was completed with the fear of being in a tight place and unable to move relieved. Misperception produces fear. To use the metaphor of the rope and the snake: although all the time there is only a rope on the ground, if you pass by it at dusk, you misperceive it as a snake and get frightened. The rope never turned into a snake. It was only misperception that produced fear.[4]

Factors Affecting Perception

Perception is seeing the things you don't register, but there are many factors to take into account that may affect the extent of perception.

Experience

This is based on past events, the things we have been exposed to, the things we have seen and done. Every encounter we have is an opportunity to learn and gain. One episode of an event brings knowledge, understanding, and know-how. Experience, in other words, may be defined as the building blocks of future recognizable events.

Education and Training

A sequence of events will be more easily and quickly recognized, analyzed, and acted upon by a person of higher education, or at least one educated and trained in the particular field of occupation.

Bias and Prejudice

Preconceived thoughts and feelings will render a person incapable of seeing other people, events, or places in any manner other than what they want to see or what they think they have seen. Prejudice will color a person's perception.

Pain and Physical State of the Observer

If an observer is experiencing pain, a fever, or other abnormality, such as a common cold, her concentration will be more focused on her personal comfort rather than her surroundings.

Complacency

Routine and sameness encourage mistakes in observation and reaction and the easy overlooking of subtle changes. For example, it has been established that the most vulnerable time for a victim to be ambushed is in the early morning hours while he is in transit and at close range, usually between his home and office or within three and a half miles from home. There are a few reasons for this.

- Primarily because the victim has a set routine, leaving home at nearly the same time every morning, there may not be many people in the area to witness the attack, and there are usually a set number of streets or roadways (usually only one or two, at most) by which the person can leave his residence.
- Being close to home in a familiar and "friendly" environment, the potential victim has a tendency to become complacent, "dropping his guard."
- Other factors include:

    1. Distance from something—a close object is more likely to attract attention than something occurring a long way away.
    2. The sound and noise relating to previous experience (a loud noise will more readily attract and be identified).

3. Movement (a moving object is more readily seen than a stationary one).
4. Size (obviously a large object will be seen quicker than a small object).
5. Color (contrasting colors will attract the eye).
6. Lighting (or lack of).
7. Education, experience, social standing … distractions.

The Fight, Flight, or Freeze Reflex

If the brain is "taken by surprise," the most common reaction will be to freeze. When a potential danger is identified, the brain analyzes the degree of danger and prepares the body for fight, flight, or freezing in place; metaphorically turning on a switch releasing a flow of adrenalin into the blood stream and to the involuntary (autonomic) nervous system, preparing the body for energetic, emergency action by increasing blood flow to the heart and opening blood vessels. The autonomic nervous system controls vital functions like blood pressure, heart rate, breathing, sweating, and other responses beyond our direct conscious control. The medical designation for adrenalin is *epinephrine*. Epinephrine stimulates the heart and is the most effective drug for countering the effects of anaphylactic shock resulting from allergies, for example to bee stings or spider bites.

When a pedestrian sees a bus or car barreling toward her, the simple thing to avoid the danger is to get out of the way and let it pass. The body stands still while the brain processes the degree of danger. Seeing, recognizing, and reacting take split seconds and in the meantime the threat is approaching closer and closer. If the brain does not command the body to move, the person is in immediate danger of being struck by the speeding vehicle. It is at this instant, while the brain is processing the danger, that all thoughts of getting out of the way for safety's sake abandon the person and she is left stuck in her tracks. The brain doesn't scream "Move! Get out of the way!" And there is no instruction given of how to get out of the way. The thought process completely stops, thus the body cannot move to safety. This is the "freeze" aspect so common in many circumstances. The brain fails, leaving the victim to watch death approaching. This commonality has come to be recognized as the "deer in headlights" syndrome.

When a deer is standing in a dark road, or nearby, and the headlights of a car strike him, he will be momentarily blinded and continue to stand in place, as though his brain is assessing the danger, long enough for him to be shot. Naturally it is illegal to hunt a deer in this manner, but that doesn't disqualify illegal hunting or shooting. A deer's first natural instinctive reaction to potential danger is to stand perfectly still and blend into the surroundings, which in the case of a nearby predator isn't always the most suitable way to survive. The "deer in headlights" has become a common definition of the human condition of freezing when momentarily stunned by being confronted with an unexpected threat such as a car bearing down on her or she is confronted by another immediate threat to her life and safety.

Rational and Irrational Fears

Rational Fears

A *rational* fear is a fear grounded in the reality of sensing and recognizing certain injury or death. Common sense tells us that situations in which the knowledge that death or injury is unavoidable and imminent are rational fears. These could include: being stalked by a person intent on committing harm; sitting in your car in the middle of an intersection, with another car approaching from the side, seeing it can't stop, and knowing it will forcibly broadside you; being trapped in a house on fire; falling into a flooding and raging stream.

Irrational Fears

An *irrational* fear is more commonly described as a phobia. It is an unreasoning response to an imagined consequence, or on encountering an object or situation that makes the sufferer feel discomfort or anxiety. Examples include: people who have minimal contact experience with the police yet have a fear of seeing an officer approaching; a person fearful of going to a doctor because he might be told he has stage three cancer.

> **BOX 4.1 "WHISTLING PAST THE GRAVEYARD"**
>
> There seem to be two meanings for this idiom, both dependent on the same metaphoric setting and action.
>
> The first connotes a situation in which a person does something (whistling, maybe?) to make a show—to others, or even more commonly, to oneself—of bravery, or at least nonchalance, in the face of danger or difficulties [pretending to be brave in a frightening environment—of course pretending may lead to an overreaction].
>
> The second meaning describes an individual who is genuinely confident and cheerful while in pursuit of a course of action at the same time blithely oblivious to the real risks involved—i.e., clueless.[5]
>
> If walking through a grave yard at night, do not be afraid of the people buried there, they are all dead; the ones to fear are the living who may have some evil purpose. Whistling will alert them to your presence and perhaps frighten them and the spirits away.
>
> Warning: It is said that whistling in a graveyard summons the devil and evil spirits.
>
> Whistling past the graveyard has to do not just with confronting fear, but also directly with death and one's proximity to it. Whistling is obviously a thing most people do when they're feeling good, so when passing the graveyard whistling, the connotation is definitely one of summoning courage to dispel the fear, but also perhaps more profoundly that one is still above ground, so you whistle as a diversion to ignore that you will yourself be dead someday, and also as expression of life, that you are in fact still alive by chance or by luck for the time being ...

Bravery and Cowardice

The Greek philosopher, teacher, and scholar, Aristotle, defined everything as having polar opposites and a center. He talked about courage, saying that on a linear scale, foolhardiness is at one end and cowardice at the extreme opposite end. In the middle of the poles is a dividing line with courage on one side and fear on the other. The degree of fear and courage is measured by how far from the polar opposites they lie (see Figure 4.1).

Foolhardiness means reckless behavior going beyond reasonable common sense; taking unnecessary risks having potential consequences of serious injury and death; whereas

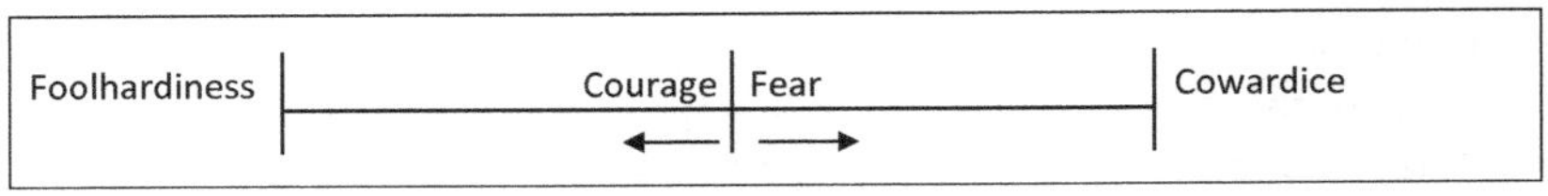

FIGURE 4.1 Aristotle's scale of fear and courage

cowardice is a total lack of courage bordering on ravages of complete terror and panic. The scale glides back and forth from one side of the middle line to the other depending on a person's perspective of the threat to his safety.

Cowardice is weakness, fearfulness, cringing at any element of danger. Whereas foolhardiness is taking unmitigated risks; placing oneself in intentional danger with no thought of the potential life-threatening consequences. Foolhardiness is walking a tight rope over Niagara Falls or attempting to fly over the Grand Canyon on a motorcycle.

Being fearless in the face of danger can be exhilarating. One could say a "thrill-seeker" or "adrenalin junky" is foolhardy, living for the "rush" of adrenalin when participating in an activity that very easily could result in serious bodily injury or death. That type of activity is a planned practice having no direct connection to the sudden startle or fright that engages the fight, flight, or freeze hormones.[6]

Cowardice is an over-produced reaction to the adrenalin "dump." Every unseen, unrecognized, or unusual sound, smell, or action is a precursor to hormonal activity, sending the recipient to retreat from the source of the fear, fear of the abstract unknown.

The conscious brain scans all emotions, sounds, etc. and sends that information to the adrenal glands, which release adrenalin calling the body into high alert for immediate action or reaction. Final judgment results by putting your body on high alert, ready to either fight or flee from an unknown enemy. The combined emotions of being tense, anxiety ridden, alert, stressed, uneasy, ready to run or fight to the death with the unknown, is felt as fear.

> You have to really be courageous about your instincts and your ideas. Otherwise you'll just knuckle under. People say to me all the time, you have no fear. I laugh out loud at that. I tell them, No, that's not true. I have as much fear as most. You have to have fear in order to have courage. I'm a courageous person because fear will never be the leader. I truly believe that everything that we do and everyone that we meet is put in our path for a purpose. There are no accidents; we're all teachers—if we're willing to pay attention to the lessons we learn, trust our positive instincts and not be afraid to take risks.[7]

When you pay attention to boredom it gets unbelievably interesting. Do just once what others say you can't do, and you will never pay attention to their limitations again.

> Everyone is a coward. People you call brave are just façades. Deep down they are as cowardly as anyone else; rather, more cowardly. Just to hide that cowardliness they have created a bravery around them, and sometimes they act in such a way that everyone knows they are not cowards. Their bravery is just a screen. How can man be brave? … Because death is there. How can man be brave? … Because man is just a leaf in the winds. How can the leaf help trembling? When the wind blows the leaf

> will tremble. But you never say to the leaf: "You are a coward." You only say that the leaf is alive. So when you tremble and fear takes grip of you, you are a leaf in the wind. Beautiful? Why create a problem out of it? But society has created problems out of everything.[8]

Even in combat there are those who will fold with fear when under pressure. Here are two true examples witnessed by a civilian contractor in Baghdad.

> A Diplomatic Security Special Agent (DSA) was riding along with our tactical Support Team when the vehicle was hit with an EFP (Explosively Formed Projectile). The driver lost an arm, but for the most part everyone in the back of the vehicle was good to go, just a little shaken up. When the crossover and EVAC started happening, it took a female IA (Intelligence Analyst) who also happened to be riding with us that day grabbing the DS Agent in attempts to drag him out from the fetal position he adopted shortly after the explosion to another vehicle. The driver with the blown off arm had already started applying his own tourniquet before the rest of the team got to him.
>
> I remember being on a recovery operation January 23rd 2007, where I was part of a Tactical Support Team in which we, along with other teams, went out to recover the bodies of four of our team mates killed when their helicopter was shot down in Baghdad. This neighborhood was also known as having a huge Sunni Extremist stronghold. The helicopter was shot down because they went in selflessly as a Quick Reaction Force and Extraction element for a protective detail that was under attack on venue. Later that night, as we were decompressing I found out that one of our team mates refused to get out of the vehicle as part of the fire team to maneuver through the neighborhood to get to the actual crash site. He was then labeled a coward and shunned from running any more missions with us. What he should have done is accept responsibility for his actions. By remaining in theater, he became a liability to everyone. He was later transferred to another position in which he was less likely to be involved in anything in which he could be in danger and a liability to others. I learned later on, he screwed up that position as well.[9]

"A hero dies only once, a coward dies a thousand deaths"—We don't do it for ourselves, we do it for the person on the right and left of us. That is unselfish bravery and courage.

> You're always going to be afraid, even you learn to hide it. Fear is like a companion, a constant companion—always there. But that's OK because fear can bring us together. Fear can bring you home. I'm going to leave you something just so you always remember. Fear makes companions of us all … I know you're afraid, but being afraid is all right. Didn't anybody ever tell you, fear is a superpower? Fear can make you faster and cleverer and stronger. If you're very wise and very strong, fear doesn't have to make you cruel or cowardly, fear can make you kind.
>
> Doctor Who[10]

The Three Hundred

Looking at bravery and courage through the ages, no example is better found than the battle of Thermopylae (Greece versus the Persian army of Xerxes 480 BCE). Three hundred dedicated brave Greek soldiers stood against an army believed to number between 200,000 and two million, facing overwhelming odds of death over survival. When King Leonidas of Sparta arrived at Thermopylae, he said to his comrades in arms, "They say that the barbarian has come near and is coming on while we are wasting time. Truth, soon we shall either kill the barbarians or else we are bound to be killed ourselves." When his soldiers complained that the barbarians were firing so many arrows at them that the sun was blocked out, Leonidas replied, "Won't it be nice, then, if we shall have shade in which to fight them?"

Another commented fearfully that the barbarians were near, he said, "Then we also are near to them." When a comrade asked, "Leonidas, are you here to take such a hazardous risk with so few men against so many?" "If you men think that I rely on numbers, then all Greece is not sufficient, for it is but a small fraction of their numbers; but if on men's valor, then this number will do." When another man remarked the same thing he said, "In truth, I am taking many if they are all to be slain."[11]

Summary

Fear can be controlled or at least minimized by using mind tricks and maintaining physiological control of your body. One method is to breathe deeply at a controlled pace to slow respiration and focus the brain on what is causing fear. Another method is to address or confront the fear and acknowledge it while assessing the degree of immediate danger and react according to the forces necessary to resolve the circumstances causing fear.

There are rational and irrational fears. Rational fears are when there is a likelihood of immediate danger, perhaps life threatening or having the potential for great bodily harm. Irrational fears are less serious but very real to the person experiencing them. These include such things as fear of spiders, regardless of the size; fear of leaving a house or apartment; fear of germs or being trapped in an enclosed space. Irrational fears are referred to as phobias.

Fear causes a response in the brain by releasing adrenalin, forcing many physiological changes in the body resulting in a fight, flight, or freeze response. There is a middle line between cowardice and foolhardiness with people fluctuating back and forth on the scale of bravery and fear depending on the perspective of the danger. Pretending to be brave in a fearsome circumstance may be foolhardy if the danger element is immediate and uncontrollable.

Review Questions

1. Explain the physiological responses to fear.
2. List and describe ways to overcome fear.
3. Diagram and explain the continuum of fear and courage.
4. Why does giving the illusion of bravery often result in unfavorable consequences such as overreaction?
5. Why should someone consider the metaphor of whistling while walking past a graveyard as a compensation for being fearful?

Notes

1 www.greatthoughtstreasury.com/author/dale-carnegie-originally-spelled-dale-carnegey. Retrieved July 3, 2015.
2 MacDonald, Cheryl A. Dr. http://healthpsychology.org/emotional-effects-of-terror/. Retrieved May 5, 2017.
3 Mobbs, Dean, et.al. *Tarantulas Help Parse Fear in the Human Brain;* Medical Research Council— Cognition and Brain Sciences Unit, Cambridge, UK; Proceedings of the National Academy of Sciences; as appeared in the *Los Angeles Times*, November 10, 2010.
4 www.notguiltybyyaani.com/chapters/book-ch9.html. Retrieved August 14, 2015.
5 http://disappearingidioms.com/whistling-past-the-graveyard/. Retrieved February 14, 2019.
6 www.ncbi.nlm.nih.gov/pmc/articles/PMC1299209/. Retrieved March 28 2017.
7 Quoted from a high-ranking martial artist friend of the author.
8 "Empty Your Cup" http://purifymind.com/EmptyCup.htm. Retrieved October 5, 2010.
9 Reply to an anonymous fear survey.
10 www.tvfanatic.com/quotes/shows/doctor-who/episodes/listen/. Retrieved March 5, 2019.
11 www.thoughtco.com/sayings-of-leonidas-116333. Retrieved September 18, 2018.

5

FEARFUL SOCIETIES OF THE PAST

A Look Back

Fear can be traced back to the earliest days of Man huddled around a fire or in a cave. The long, dark nights brought anxiety and fear to the first humanoids approximately eight million years ago.

> Darkness covered the earth. Nighttime had fallen and the human-like beings huddled together for warmth from the winds driving night sounds across the veldt. The roar of a saber tooth tiger echoed across the scrub brush and tall grass while a pack of hyenas on the opposite side of the tall canyon ripped at the carcass of a freshly fallen antelope. The nearly human creatures dared not move, as their eyes tried to pierce the darkness, looking out for the approach of a stealthy cat or other creature that was sure to pounce on them. Fear created a smell of its own. The only thought, if it could be construed as a thought, through the long fearful night, was survival until the next daylight. The Clan realized by instinct that their safety was dependent upon staying together as a group, forming a circle with the strongest and bravest on the outside ring. It was the early dawn of Man. During those first millenniums Man's only instincts and needs for survival were for food, clothing (for warmth), shelter, sex, and security.[1]

It was only with the coming of the morning sun, refreshing the day, that the nighttime terrors went away. In their primitive human minds there was no distinction between day and night except for the darkness when the "ball of fire" hanging over them moved across the sky until it was swallowed by unknown forces to be reborn after many hours of anxiety, fear, terror, and panic each time the roar of a beast sounded. They had no cognition of the mystery of why or how the sun rose and fell; all they understood was that darkness brought unseen causes of fear. To explain the unexplainable, Man eventually created "Gods" to help him understand.

> FEAR was the first thing on Earth to make the Gods.[2]

> The origin behind myths and religion is human terror of annihilation. Human societies invented mythology and religion in order to militate against people's fear of living a mortal life. People fear time as a destroyer of human happiness, human beings, and human societies.[3]

Gods of the Battlefield

Among the created gods were Deimos and Phobos. Deimos (or Deimus) was the God (*daimon*) of fear, dread, and terror, and his twin brother Phobos (Phobus) of panic, fear, flight, and battlefield rout. They would accompany their father Ares into battle, who would drive his chariot and spread fear in his wake. They were also the sons of Aphrodite, goddess of love, and represented the fear of loss.[4] Today we think of Deimos as Demons and Phobos as Phobias.

Deimos (fear) and Phobos (flight) are the gods of the battlefield and war.

> Among the images engraved on the shield of Herakles: In the centre was Phobos (Fear) worked in adamant, unspeakable, staring backwards with eyes that glowed with fire. His mouth was full of teeth in a white row, fearful and daunting, and upon his grim brow hovered frightful Eris (Strife) who arrays the throng of men … Upon the shield Proioxis (Pursuit) and Palioxis (Flight) were wrought, and Homados (Tumult), and Phobos (Panic), and Androktasie (Slaughter). Eris (Strife) also, and Kydoimos (Uproar) were hurrying about, and deadly Ker (Fate).
>
> And on the shield stood the fleet-footed horses of grim Ares made gold, and deadly Ares the spoil-winner himself. He held a spear in his hands and was urging on the footmen: he was red with blood as if he were slaying living men, and he stood in his chariot. Beside him stood Deimos (Fear) and Phobos (Flight), eager to plunge amidst the fighting men.[5]

The Cult of Assassins

Present-day terror in the Middle East has its roots in the eleventh and twelfth centuries (approximately 1090–1273) with the establishment of a shadowy Muslim offshoot known as Nizari Ismailis, founded at Altamont, Persia by a mysterious figure, Hasan al Sabbah. They "came to be known by the chilling sobriquet, the 'Order of Assassins.'"[6] They struck fear and daggers into the hearts of their political enemies.

FIGURE 5.1 An assassin

The trained and loyal assassins worked for and won acceptance into the courts of the enemies of the Ismailis. They often remained for years in trusted positions of their intended victim until receiving orders to commit the assassination. Upon getting close enough, the fateful dagger was drawn and shoved into the chest of the intended victim, whether he was fully guarded or not. Each assassin was personally selected for his willingness to die for the objective of striking fear in the greater rulers. All the assassins knew they were going to be captured and killed. Each mission was suicidal. Centuries later, terrorist groups use suicide bombers to spread fear and terror.

The Reign of Terror—The Great Grandmother of Terrorism

"Guillotine! … Guillotine!" Madame Defarge,[7] her hands and fingers moving deftly through her knitting, stands in the front row of the crowd chanting for the beheading of another condemned aristocrat. "Guillotine! Guillotine!" The chant grows louder and more rhythmic, Madame Defarge's stitching needles, clicking in time with the chant, secretly knit a register of the names of the revolution's intended victims. The setting is Paris; the time is the middle of the French Revolution (1789–1799). It was a bloody, chaotic time.

> It was the best of times, it was the worst of times, it was the age of wisdom, it was the age of foolishness, it was the epoch of belief, it was the epoch of incredulity, it was the season of Light, it was the season of Darkness, it was the spring of hope, it was the winter of despair, we had everything before us, we had nothing before us, we were all going direct to Heaven, we were all going direct the other way—in short, the period was so far like the present period, that some of its noisiest authorities insisted on its being received, for good or for evil, in the superlative degree of comparison only.[8]

"J'accuse!"[9]—Bitterly Denounce

It was a time that came to be known as the "Great Fear" or, more commonly, the "Reign of Terror."

> During the Reign of Terror of the mid-1790s, thousands of "enemies of the French revolution" (including King Louis XVI and his wife Marie Antoinette) met their end by the guillotine's blade. Some members of the public initially complained that the machine was too quick and clinical, but before long the process had evolved into high entertainment. People came to the Place de la Revolution in droves to watch the guillotine do its grisly work, and the machine was honored in countless songs, jokes and poems.
>
> Spectators could buy souvenirs, read a program listing the names of the victims, or even grab a quick bite to eat at a nearby restaurant called "Cabaret de la Guillotine." Some people attended on a daily basis, most famously the "Tricoteuses," a group of morbid women who supposedly sat beside the scaffold and knitted in between beheadings. The theater even extended to the condemned. Many people offered sarcastic quips or defiant last words before being executed, and others danced their way up the steps of the scaffold.[10]

The reign of terror in those muddled times is the etymology of the modern-day term, terrorism. The accepted estimated number of people beheaded is set at 16,594. The era was

FIGURE 5.2 Execution of Louis XVI
Source: www.wikiwand.com/en/Execution_of_Louis_XVI

frightening due to no one knowing when they would be denounced as "an enemy of the revolution."

"*J'accuse!*" became a common cry that unto itself brought fear to anyone suspected of being in any way connected to or associated with nobility. A mere accusation was accepted as proof and many innocent people were accused, thrown into prison, and led to an appointment with "Madame Guillotine."

Maximilien de Robespierre, a leading member of the Jacobin party, came into power on the back of common street violence and the revolutionists. He called for the death of Louis XVI claiming, "Louis must die in order for the revolution to live." For the first time in history terror became an official government policy, with the stated aim to use violence in order to achieve a higher political goal. Unlike the (modern) meaning of "terrorists as people who use violence against a government," the terrorists *were* the government. The terror was legal, having been voted for by the National Convention (the new government formed by Robespierre).[11]

Robespierre "spoke of conspiracies against the Revolution, of the threats to 'the patrie in danger' and the need for extreme measures, he voiced the fears of many at the time that France was about to be overwhelmed by foreign and internal enemies."[12] He had good reason to hold that belief because of concurrent economic chaos, and ongoing foreign wars against Austria, Prussia, Britain, Holland, and Spain (very similar to contemporary fears of the government). On July 27, 1794 Robespierre was arrested and executed the next day, as the reign of terror began to wind down.

Invisible Empire of the South—"White Supremacy Terror"

Two years after the American Confederate army surrendered to Federal forces at Appomattox, Virginia, in 1865, a social club consisting primarily of former confederate soldiers was created. The founders took the name of the organization as Ku Klux Klan, with the first two words, *Ku Klux* coming from the Greek word, *kyklos*, meaning circle. A year later offshoots of the club met in a unifying organizational meeting and elected a former confederate general, Nathan Bedford Forrest, as the leader or, as the position was called, *Grand Wizard.*

The Klan professed to be a Christian organization with an agenda to reverse the governmental policy of reconstruction, which allowed blacks to vote and hold office, and restore white supremacy in the south. Their violent methods were anything but Christian. Spreading fear and violent death was the Klan's chosen *modus operandi* (method of operation) or *actus reus* (criminal act) in their campaign of spreading terror throughout the Southern states. Wearing white hoods over their faces and long white robes over their clothing, a group of Klansmen on horseback would suddenly appear in the dark of night at their selected victim's home, grab the frightened man of the black family and hang him from the nearest tree, often setting his house afire with the family inside.

Hangings and raids became common place with the Klan's calling card of leaving a large burning cross in the yard. In the actual words of former slaves who were witness to the KKK terror:[13]

> They was terrible dangerous. They wear long gowns, touch the ground. They ride horses through the town at night and if they find a Negro that tries to get nervy or have a little bit for himself, they lash him nearly to death and gag him and leave him to do the best he can. Sometimes they put sticks in the top of the tall thing they wear and then put an extra head up there with scary eyes and great big mouth. Then they stick it clear up in the air to scare the poor Negros to death .... Them was bad times, them was bad times. I know folks think the books tell the truth but they sure don't. Us poor niggers had to take it all.
>
> I never did have no truck with de Ku Kluckers, but I had to step mighty high to keep outen it. De sure 'nough Kluxes never did bother around us 'cause we minded our own business and never give no trouble.

FIGURE 5.3 Ku Klux Klan
Source: www.wikiwand.com/en/Ku_Klux_Klan

> Right after de war de Ku Klux got after de colored folks. Dey would come to our houses and scare us most to death. Dey would take some of de niggers out and whip 'em and dose dat dey didn't whip dey tied up by dere fingers and toes. Dose Ku Klux would come to our windows at night and say: "Your time ain't long a-comin'." De Ku Klux got so bad dat dey would even get us in de daytime. Dey kept dis up till some folks fom de North come down and put a stop to it.

The name Ku Klux Klan is synonymous with the words "White Supremacy Terror." The Klan is still active in the United States and continues to spew hatred and fear.

Spanish Inquisition

The Reign of Terror and the founding of the Ku Klux Klan are mere moments in the long history of hate and terror. There was a time when "The Church" ruled the world, being more powerful than all the kings and emperors. An era of fear and denunciation in the name of religion is the Spanish Inquisition.

> All crimes and all vices are contained in this one word, *Inquisition*: Murder, robbery, arson, outrage, torture, treachery, deceit, hypocrisy, cupidity, holiness. No other word in all languages is so hateful as this one that owes its abhorrent preeminence to its association with the Roman Church.[14]

> To think of the Spanish Inquisition is to associate thoughts with terror, horror, and religious persecution. This era of severe censorship, paranoia, torture, autos-da-fé, death, and the general persecution of heretics, or in other words pretty much anybody who deliberately disagreed with the principles of the Catholic church, began in 1478 and lasted until 1834.[15]

To be accused of "heresy,"[16] or simply being Jewish, Muslim, or not a Catholic Christian could, and most often would, land a person in the hands of the inquisitors who would demand a conversion to Catholicism. If the arrested person did not convert, he would be subjected to horrific torture and death. "Anyone who attempts to construe a personal view of God which conflicts with Church dogma must be burned without pity" (Pope Innocent III).

> Heresy has existed in every age, but during the 12th century, the Catholic Church took unprecedented action against it. As the power of the Catholic Church increased in Europe, dissenting voices of other Christian groups became more troublesome. Pope Alexander III (1162–63) encouraged informers, so the church could discover evidence of heresy. In 1184 Pope Lucius III issued a decree that a convicted heretic was to be handed over to secular authorities for punishment. Over the next several decades, the church increased the severity of punishment for heresy, ultimately making it a capital offense under Pope Gregory IX. During this time, the Dominicans became the principal agents of the Inquisition; a special court given authority to judge intentions as well as actions. When heresy was suspected in a village, an inquisitor was sent to preach a sermon calling for the villagers to come forward with reports of heresy. This was a "general inquisition" that included a period of grace for anyone who would confess. This was followed by a "special inquisition" that might include coercion, false witnesses, and torture to obtain a "confession." Those identified as heretics were then ordered to do penance, which might consist of mandatory church attendance, pilgrimage to a shrine, loss of property, or imprisonment. Heretics who refused to repent were sentenced to death.[17]

> Firstly when the faith is in question, there must be no delay; but at the slightest suspicion, rigorous measures must be resorted to with all speed. Secondly, no consideration is to be shown to any prince or prelate, however high his station. Thirdly, extreme severity is rather to be exercised against those who attempt to shield themselves under the protection of any potentate, and fourthly, no man must lower himself by showing toleration toward heretics of any kind.[18]

Means of Torture

Torture was used to obtain confessions from those accused of heresy and not following the dictates and premise of the Roman Catholic Church. The official stance was that torture was not meant for punishment but to obtain converts and confessions. Like the days of the French Revolution, anyone could and would denounce anyone without proof. The mere accusation was sufficient to cause an arrest, trial, and torture.

> Denunciations were anonymous. Defendants had no way of knowing the identity of their accusers. This was one of the points most criticized by those who opposed the Inquisition. In practice, false denunciations were frequent, resulting from envy or personal resentments. Many denunciations were for absolutely insignificant reasons. The Inquisition stimulated fear and distrust among neighbors, and denunciations among relatives were not uncommon.[19]

When a person was forced to confess to his sins, he was also forced to name any and all others the Inquisition could accuse. This led to many innocent people being implicated, arrested, tortured, and killed.

Types of torture varied, and a very small percentage of those accused escaped the torture chambers. Often the accused would be walked through the hall with all the torture devices and allowed to see them being used, then thrown into an underground dungeon in darkness, perhaps chained to the wall in a standing or sitting position, where he would remain for years, starving and slowly going insane.

> The dungeons were situated underground, so that the outcries of the subject might not reach other parts of the building. In some medieval cells, the inauspicious were bound in stocks or chains, unable to move about and forced to sleep standing up or on the ground. In some cases there was no light or ventilation, inmates were generally starved and kept in solitary confinement in the dark and allowed no contact with the outside world, including that of their own family.[20]

Often just viewing the torture was enough to convince a person to confess, but it was taken as not being sincere so the individual was subjected to the dungeon and the torture devices from which few survived. Similar to modern day torture by waterboarding, the first order of torture was forcing the victim to swallow so much water it caused a stomach swelling. The victim was also starved and hot burning coals were placed on his naked body. A method called *strappado* was a form of torture with the hands tied behind the back and lifted off the ground by a pulley with a rope attached to the wrists. This would cause the shoulders to be painfully dislocated or pulled from their sockets. In addition, the prisoner would be dropped nearly to the floor then roughly jerked back up, imposing more pain.

One of the most notorious devices of torture was the rack. The accused heretic was placed on a wooden table or platform with rollers on each end. His hands and feet were tied to the rollers. The Inquisitor would turn a crank handle, tightening the bindings, stretching the victim until his limbs were pulled from their sockets or ripped off. While the helpless victim was tied to the rack or hanging from the *strappado* it was not uncommon for the torturer to use white hot pincers, thumb screws, and other painful devices designed to maim, injure, or mutilate the torso, hands, and feet.

During the centuries of the Inquisition, terror reigned supreme, with the only rival of wholesale terror in the degree of terror being the French Revolution. Until …

Salem Witch Hunts (1692–1693)

On a far lesser scale, fear was abundant in the Puritan colonial village of Salem, Massachusetts. The Puritan strict interpretation of the bible and the intrinsic belief in sin, acts of the devil, and witchcraft led to hysterical fear beyond reason. With heavy reliance on "Biblical Law" and superstition, the people of the early village of Salem believed witchcraft was a work of the Devil affecting one then several other girls, leading to a "witch hunt" spurred by unsupported accusations and coerced confessions. During the height of the frenzy everyone was terrified that they might be the next accused victim.

> From June through September of 1692, nineteen men and women, all having been convicted of witchcraft, were carted to Gallows Hill, a barren slope near Salem Village, for hanging. A man of over eighty years was pressed to death under heavy stones for refusing to submit to a trial on witchcraft charges. Hundreds of others faced accusations of witchcraft. Dozens languished in jail for months without trials.[21]

What was it in Puritan faith that allowed the denouncing of good people and accusing them of being "servants of the devil"?

> The Puritans believed that the Bible was God's one true law and it provided all that was needed for life and living. They felt the Church of England had become a product of political struggle and manmade doctrines and was beyond redemption. Hundreds of successful thriving communities had been formed by the 1700s. Since God and the Bible was at the forefront of everything they did it was the motivating ideal that worked both for them and against them. When the "Pilgrims" landed they had no laws other than the law of the Bible.
>
> Their interpretation of scriptures was harsh and unyielding. Any deviation from the strict Puritan life was met with disapproval and punishment. The elders of the church were also the political leaders, so any infraction was both social and religious and there was no margin of error.
>
> The devil was behind every misdeed and every misunderstood event even natural ones like rain, wind and snow storms. Human misdeeds, jealousy, greed and differences could be justified and blamed on witchcraft since proving witchcraft was impossible, it was simply a matter of suggesting its presence. Once the seed of fear was planted, the old woman who used plants to heal was "a witch" the young woman who smiled at your intended was "a witch" The neighbor whose stream you coveted was "a witch." When the young girls Betty Paris and Abigail Williams fell ill and the village doctor could not find or cure the problem, he proclaimed they were "bewitched" and the hunt was on.[22]

The law in colonial Massachusetts was conceived from interpretations of scripture in the bible and influenced by Calvinism. Reformation leader John Calvin "held that human beings were innately sinful—utterly depraved by inheriting the original sin of Adam and Eve, the biblical parents of the human race." Believing in "predestination" in Calvinism meant

> God decided who would be saved or damned before the beginning of history—and that this decision would not be affected by how human beings behaved during their lives .... Its power to comfort and reassure troubled souls arose from its wider message that, beyond preordaining the eternal fates of men and women, God had a plan for all of human history—that every event in the lives of individuals and nations somehow tended toward an ultimate triumph of good over evil, order over disorder, Christ over Satan. In other words, Calvin (and his many followers among groups like the Puritans) saw human history as an unfolding cosmic drama in which every person had a predestined role to play. True, men and women had no free will, but they had the assurance that their existence—indeed, their every action—was MEANINGFUL and that their strivings and sufferings in the present would ultimately produce a future of perfect peace and security—a kind of heaven on earth .... They strove to lead godly and disciplined lives—but not because they hoped that such righteous behavior would earn them salvation. Instead they believed that their very ability to master their evil inclinations provided some evidence that they ranked among the elect of saints. In other words, the Puritans did not regard leading a godly, moral life as the CAUSE of a person's salvation, but rather as an encouraging sign of the EFFECT of being chosen by God to enjoy eternal bliss in heaven.[23]

The strictness of Puritanical religious law and the power of the Church over State affairs eventually led to the premise of Separation of Church and State in the American Constitution.

Comparing Puritan and Sharia Law

When the Puritans established the Massachusetts colony there were no laws except what they created following the bible and, as "God Fearing" citizens, they believed in strict adherence to "God's Words." The 1641 capital laws of New England were collected and crafted from the texts of the bible. The laws are transcribed here in the old English and from these it can be easily understood why fear of being accused of a crime could have serious consequences.

Complicating matters, the accused was required to prove his innocence. In the modern United States there is an assumption of innocence until proven guilty beyond a reasonable doubt. Sharia law expects the requirement of the accused proving his innocence, with no assumption of innocence until proven guilty beyond a reasonable doubt. It is practically impossible to prove innocence. For example, how can a man prove he "had no sex with that woman"? In Sharia law, adultery is punishable by stoning the accused woman to death. How can she prove she did not have sex with a man?

Capitall Lawes, Established within the Jurisdiction of Massachusetts[24]

1. If any man after legall conviction, shall have or worship any other God, but the Lord God, he shall be put to death. Deut. 13. 6, &c. and 17. 2. &c. Exodus 22. 20.

2. If any man or woman be a Witch, that is, hath or consulteth with a familiar spirit, they shall be put to death. Exod. 22. 18. Lev. 20. 27. Deut. 18. 10, 11.
3. If any person shall blaspheme the Name of God the Father, Sonne, or Holy Ghost, with direct, expresse, presumptuous, or high-handed blasphemy, or shall curse God in the like manner, he shall be put to death. Lev. 24. 15, 16.
4. If any person shall commit any wilfull murther, which is manslaughter, committed upon premeditate malice, hatred, or cruelty, not in a mans necessary and just defence, nor by meer casulatie, against his will; he shall be put to death. Exod. 21. 12, 13, 14. Num. 35. 30, 31.
5. If any person slayeth another suddenly in his anger, or cruelty of passion, he shall be put to death. Num. 35. 20, 21. Lev. 24. 17.
6. If any person shall slay another through guile, either by poysonings, or other such divilish practice; he shall be put to death. Exod. 21. 14.
7. If a man or woman shall lye with any beast, or bruit creature, by carnall copulation, they shall surely be put to death; and the beast shall be slaine, and buried. Lev. 20. 15, 16. 49.
8. If a man lyeth with mankinde, as he lyeth with a woman, both of them have committed abomination, they both shall surely be put to death. Lev. 20. 13.
9. If any person committeth adultery with a maried, or espoused wife, the Adulterer, and the Adulteresse, shall surely be put to death. Lev. 20. 10. and 16. 20. Deut. 22. 23, 24.
10. If any man shall unlawfully have carnall copulation with any woman-childe under ten yeares old, either with, or without her consent, he shall be put to death.
11. If any man shall forcibly, and without consent, ravish any maid or woman that is lawfully married or contracted, he shall be put to death. Deut. 22. 25. &c.
12. If any man shall ravish any maid or single woman (committing carnall copulation with her by force, against her will) that is above the age of ten yeares; he shall be either punished with death, or with some other grievous punishment, according to circumstances, at the discretion of the Judges: and this Law to continue till the Court take further order.
13. If any man stealeth a man, or man-kinde, he shall surely be put to death. Exod 21. 16.
14. If any man rise up by false witnesse wittingly, and of purpose to take away any mans life, he shall be put to death. Deut. 19. 16. 18, 19.
15. If any man shall conspire, or attempt any invasion, insurrection, or publick rebellion against our Common-wealth, or shall indeavour to surprize any Towne or Townes, Fort or Forts therein: or shall treacherously, or perfidiously attempt the alteration and subversion of our frame of pollity, or government fundamentally, he shall be put to death. Num. 16. 2 Sam. 3. & 18. & 20. 50

Sharia Law

In 2014 a very conservative and radical form of Islamism emerged from the chaos of the Middle East. One of their primary purposes was to conquer wide swaths of territory and to eventually take over the entire world under one caliphate to spread their belief of living under Islamic or "Sharia" law. Sharia law is very similar to Puritanical—Biblical law. In Sharia law the religion is the law, the tenets of Sharia are a strict interpretation of the Quran. The word "sharia" means a body of moral and religious law derived from religious prophecy, as opposed to human legislation.

- Theft is punishable by amputation of the hands (Quran 5:38).
- Criticizing or denying any part of the Quran is punishable by death.
- Criticizing Muhammad or denying that he is a prophet is punishable by death.

- Criticizing or denying Allah is punishable by death (see Allah moon god).
- A Muslim who becomes a non-Muslim is punishable by death (See Compulsion).
- A non-Muslim who leads a Muslim away from Islam is punishable by death.
- A non-Muslim man who marries a Muslim woman is punishable by death.
- A woman or girl who has been raped cannot testify in court against her rapist(s).
- Testimonies of 4 male witnesses are required to prove rape of a female (Quran 24:13).
- A woman or girl who alleges rape without producing 4 male witnesses is guilty of adultery.
- A woman or girl found guilty of adultery is punishable by death (see "Islamophobia").
- A male convicted of rape can have his conviction dismissed by marrying his victim.
- Muslim men have sexual rights to any woman/girl not wearing the Hijab (see Taharrush).
- A woman can have 1 husband, who can have up to 4 wives; Muhammad can have more.
- A man can marry an infant girl and consummate the marriage when she is 9 years old.
- Girls' clitoris should be cut (Muhammad's words, Book 41, Kitab Al-Adab, Hadith 5251).
- A man can beat his wife for insubordination (see Quran 4:34 and Religion of Peace).
- A man can unilaterally divorce his wife; a wife needs her husband's consent to divorce.
- A divorced wife loses custody of all children over 6 years of age or when they exceed it.
- A woman's testimony in court, allowed in property cases, carries ½ the weight of a man's.
- A female heir inherits half of what a male heir inherits (see Mathematics in Quran).
- A woman cannot speak alone to a man who is not her husband or relative.
- Meat to eat must come from animals that have been sacrificed to Allah - i.e., be "Halal."
- Muslims should engage in Taqiyya and lie to non-Muslims to advance Islam.[25]

In the current modern day religious-political climate, the loud voices of Islamist terrorists like ISIS, and Boko Haram's followers and similar zealots are calling for installation of and spreading of Sharia law, which is frightening and should be rejected by a free society.

The Decade of Violence

The 1960s in the United States were years of turmoil, chaos, hatred, and fear because of political assassinations, riots in the cities with looting and burning, racial discord, anti-war demonstrations, civil disobedience, and labor strikes. The consequential effects on the American public had long-range significant reverberations, still being felt today. During those tumultuous years of civil rights freedom marches, anti-Vietnam war movements, and "women's rights," the police had a difficult time maintaining peace and order, often resorting to force, which in the minds of many became "excessive force" to quell the violence. Dogs were used to control crowds; police often resorted to the use of armored vehicles along the line of military tanks to force rioters to disperse. In turn, the police became targets of those who would sanction anarchy. Police were called "pigs," certainly a show of disrespect, and the "thin blue line" became a symbol of repression.

Many times a rock or bottle would be hurled from the shadows at a passing patrol car, a police car would be dispatched to the scene of a reported racial fight in a neighborhood and upon arrival would find a vacant street only to be met with gun fire from a hidden ambusher. The times seemingly pitted law enforcement against disrupting forces fighting for a cause they deeply believed in. Society was on the verge of civil rebellion and police were the focal point, the match on a barrel of gun powder. Every police action was criticized, publicized, and condemned.

Those years exemplified the changing culture of the United States. Women and minority groups, including people of color, the poor and disenfranchised took to the streets demanding fair and equal treatment. Thus police were forced into an unpopular position of having to enforce laws while maintaining the thin line between first amendment rights of free speech, protests against the government, and freedom of assembly. Many officers suffered what, at that time, was defined as "combat fatigue" from high anxiety and accompanying stress.

Summary

From times of earliest mythology and countless gods to current social problems, fear has been a weapon to defeat enemies, ideologies, and freedom of thought. It commits racial genocide and discrimination. There are no exceptions. Everyone at some time or circumstance has been victimized by fear. It may be social or personal. It may be subtle and hidden or overt in displays of uncontrollable flight, flee, or freeze reactions, often leading to aggressive and undesired behavior.

Review Questions

1. Why did fear create the gods?
2. How did early religion and its leaders use fear to control "the people"?
3. Pick a time period, from "caveman" days to the contemporary scene, and discuss the concepts or commonalities of fear.
4. The Ku Klux Klan (KKK) and other white supremacy groups intimidate minority groups through fear. Recount a recent incident involving such a group.
5. Comparing Puritanical Law with Sharia Law, explain the similarities.
6. Examining Puritanical Law, discuss "Freedom of Religion" and why our Constitution requires separation of church and state.

Notes

1 June, Dale; Introduction to *Executive Protection*, 1st and 2nd edns; CRC Press, Boca Raton (1998 and 2008).
2 https://iperceptive.com/authors/lucretius_quotes.html. Retrieved March 5, 2019.
3 www.goodreads.com/quotes/7792109-the-origin-behind-myths-and-religion-is-human-terror-of. Retrieved March 5, 2019.
4 www.theoi.com/Daimon/Deimos.html. Retrieved July 8, 2015.
5 Ibid.
6 During the Crusades, the Muslim sect known as the Assassins tamed more powerful enemies using a shocking means: murder; www.historynet.com/holy-terror-the-rise-of-the-order-of-assassins.htm. Retrieved September 25, 2018.
7 Madame DeFarge, a fictional character in the Charles Dickens novel, *A Tale of Two Cities*.
8 Opening paragraph of *A Tale of Two Cities* by Charles Dickens. This is arguably one of the best known opening paragraphs in literature.
9 French Translation means: "Step forward to accuse the father," but has come to also mean, "I accuse."
10 www.history.com/news/history-lists/8-things-you-may-not-know-about-the-guillotine. Retrieved January 9, 2015.
11 www.historytoday.com/marisa-linton/robespierre-and-terror. Retrieved January 9, 2016.
12 Ibid.
13 *When I Was a Slave: Memoirs from the Slave Narrative Collection*; Edited by Norman R. Yetman; Published by Dover Publications, 1970. This Dover edition, first published in 2002, is a republication of selected unabridged narratives from the Slave Narrative Collection developed from interviews recorded and transcribed by the Federal Writer's Project, 1936–1938.
14 *The Horrors of the Church and Its Holy Inquisition*; www.bibliotecapleyades.net/vatican/esp_vatican29.htm. Retrieved January 13, 2016.
15 www.enforex.com/culture/spanish-inquisition.html. Retrieved January 13, 2016.
16 "Adherence to a religious opinion contrary to church dogma." A second definition is "dissent or deviation from a dominant theory, opinion, or practice." With regards to religion, any belief or

practice that goes against the official position of the church is considered heretical. www.gotquestions.org/heresy-definition.html. Retrieved January 13, 2016.

17 Ibid.

18 *The Horrors of the Church and Its Holy Inquisition*; www.bibliotecapleyades.net/vatican/esp_vatican29.htm. Retrieved January 13, 2016.

19 www.newworldencyclopedia.org/entry/Spanish_Inquisition. Retrieved January 13, 2016.

20 *The Horrors of the Church and Its Holy Inquisition*; www.bibliotecapleyades.net/vatican/esp_vatican29.htm; Retrieved January 13, 2016.

21 Linder, Douglas; *The Witchcraft Trials in Salem: A Commentary;* http://law2.umkc.edu/faculty/projects/ftrials/salem/SAL_ACCT.HTM. Retrieved January 16, 2016.

22 Dawes, James; www.answers.com/Q/The_puritans_faith_and_beliefs_about_withcraft. Retrieved January 17, 2016.

23 Heyrman, Christine Leigh. "*Puritanism and Predestination.*" *Divining America*, TeacherServe©. National Humanities Center. http://nationalhumanitiescenter.org/tserve/eighteen/ekeyinfo/puritan.htm. Retrieved January 20, 2016.

24 *The Capital Laws of New England as They Stand Now in Force in the Commonwealth, 1641, 1642*, reprinted in THE LAWS AND LIBERTIES OF MASSACHUSETTS 1641–1691 (John D. Cushing ed., 1976). Appearing in Biblical Law in America: Historical Perspectives and Potentials for Reform by John W. Welch; BRIGHAM YOUNG UNIVERSITY LAW REVIEW; 2002.

25 www.billionbibles.org/sharia/sharia-law.html. Retrieved September 22, 2019.

6

FEAR IN CONTEMPORARY SOCIETY

> Ordinary, everyday interactions between the police and the community have become tinged with fear—fear and stress that exists on both sides of the interaction. Residents are deeply afraid of crime and violence in their communities. Much of the community's fear is driven by their own experiences, and the experiences of family members and friends.
>
> Tragically, in many of the communities where residents are the most fearful of crime, they are more likely to be fearful of the people who are supposed to protect them—that is, the police. These are not so much fears about getting into physical confrontations with the police, but fears of being treated with suspicion, disrespect and derision. Such fears are found among minority residents of all socio-economic classes. Those who live in fear, learn to hate and distrust.
>
> At the same time, our police officers are also fearful. They are fearful of many of the same things the community is afraid of: the unpredictable nature of crime and violence on the streets they patrol. Officers are fearful of violence against themselves and their colleagues. Police are fearful of other things as well. They are fearful of the intense scrutiny placed on them by the media, by community activists, by the legal system, by their bosses. For many police officers, especially those working in high-crime areas, their lives have become a "good guys vs. bad guys" drama played out in the communities they serve. And these officers see so many of the latter that they tend to lose sight of the former—the good, law-abiding people who make up the vast majority of residents in even the most crime-infested communities.
>
> Fear—among police officers and community members—tends to breed mistrust, which in turn fosters stereotypes, which in turn leads to an exaggerated sense of the differences between the two groups. These situations do not take place in a vacuum. They take place in an environment shaped by the experiences, the attitudes and, yes, the fears of all the people involved.[1]

People assume stereotypes about those who do not look like them, with no facts, no evidence, and no desire to get to know the truth. Some people will judge another based on stereotypes, making broad generalizations and negative assumptions such as: black people steal and are criminals; Mexican people are lazy, illegal aliens; Muslims hate Americans and are terrorists, (or all terrorists are Muslim) hence, these people deserve to be oppressed, exploited, excluded and ostracized because they are not as "good" as the dominate group (i.e. ethnocentrism). Police officers are often stereotyped as authoritarian, donut bloated, and biased.

When discussing racism and fear, we cannot leave out ignorance. Some people are prejudiced and don't really realize it because racism is so inherent and so institutionalized and often subtle. Subtle prejudice is simple yet complex. It is as simple as locking the car doors when a black person passes in front at a traffic light and as complex as police profiling black males. Another form of overt, yet subtle, racism is a white person addressing a black man as "boy"; or black female, "girl." In slang terms saying, "my man" can have racial overtones of racism if said to a black man by a white man. The inference is the black man is owned by the white man. Statements such as these are labeled "micro aggressions." A micro aggression is "a subtle but offensive comment or action directed at a minority or other non dominant group that is often unintentional or unconsciously reinforces a stereotype."[2] Examples include: "I don't see you as black" or "I don't see color."

Ignorance about another culture/race can be a dangerous thing. Black people read white publications: *Cosmopolitan, Vanity Fair, Redbook* .... But how many white people read black publications: *Ebony, Essence, Black Entrepreneurs* ...? Not many.

An exemplifying scenario of racial fear based on ignorance is that of the wealthy white woman born, raised, and living in an all-white enclave her whole life. Her friends, associations, neighbors, and contacts are white; she has never been exposed to a black person. All her knowledge of other races comes from news media stories, rumors, and the beliefs of her friends.

One day, dressed in the latest fashion and wearing her expensive jewelry, she drives her Mercedes to the center city for an appointment. As she gets out of the car, her jewelry sparkles in the sunlight. She begins to walk toward her appointment when she sees three black youths, aged approximately 16 to 18. She clutches her purse tightly against her body and puts her hand over her diamond necklace. She then crosses the street to avoid the boys. As she walks further, she sees three Mexican boys, the same age as the black boys. She again protects her purse and re-crosses the street. At the corner she sees three white boys, again of the same age. This time she doesn't clutch her purse and cover her diamonds. Instead, she looks at the boys and keeps walking. As she rounds a corner, three boys run up behind her, knock her down, grab her purse, and take her necklace. She calls the police and says she was robbed by the three black boys. Alternatively, she describes the Mexican boys. It never occurs to her that perhaps she was mugged by the white boys. Ignorance and racism combine to form fear and factual blindness.

Cultural differences and subtleties count for a lot when starting to bridge the gap between people of different ethnic backgrounds. People fear what they do not understand. When something is feared, people usually run from it, oppress it, dominate it, or try to destroy it. It is just human nature to do so. Showing the smallest interest in someone else's culture goes a long way; people want to share what makes them unique and it helps to stamp out ignorance.

> People have more difficulty getting over fear toward members of other races than toward those of their own race, a new study shows. Arne Ohman, a professor of psychology at Karolinska Institutet in Stockholm, Sweden, wrote an article accompanying the study in *Science*: "The function of fear is to promote avoidance of potentially dangerous stimuli, and what you avoid you will not learn about," Ohman said. "Because you isolate yourself from corrective information, you may think that members (of other groups) are dangerous and evil in order to justify your fear, and in this way fear may breed interracial antagonism."[3]

In April, 2018, police officers were called to a Starbucks Coffee shop in Philadelphia to respond to the manager's concern that two black men were sitting at a table for an unreasonable time (according to the manager) without making an order. They simply sat at the table conversing while waiting for a friend. The men were arrested and taken out of the store in handcuffs. They were later released with no charges being filed.

In a strikingly similar police incident, later nicknamed "BBQ-ing while black," a white woman called police to report a black family using a barbeque grill in an area of a public park where barbeque grills were not allowed. Another example is of the black Smith College student who enjoyed her lunch in the college cafeteria then took a few minutes to relax on a couch in the student lounge. A school employee saw her and called the police. In her words the student said,

> I am blown away at the fact that I cannot even sit down and eat lunch peacefully. Today someone felt the need to call the police on me while I was sitting down reading, and eating in a common room at Smith College. This person didn't try to bring their concerns forward to me, but instead decided to call the police. I did nothing wrong, I wasn't making any noise or bothering anyone. All I did was be black. It's outrageous that some people question my being at Smith College, and my existence overall as a woman of color. I was very nervous, and had a complete meltdown after this incident. It's just wrong and uncalled for. No students of color should have to explain why they belong at prestigious white institutions. I worked my hardest to get into Smith, and I deserve to feel safe on my campus.[4]

In another event, a white woman called the police about a black man in her apartment complex because she "felt he didn't belong." She was very wrong, he actually lived there and was going to his own apartment. In a tragic, but similar instance, an off duty deputy sheriff entered what she said she thought was her apartment and saw a black man. She shot and killed him in his own apartment!

These events are only a representation of several involving a white person reporting black people "acting suspiciously." The social climate has again seemingly deteriorated to the mid-twentieth century when racial fear and violence headlined the news media.

According to a new Pew Research Center Survey, "Black and White Police officers have strikingly different views on a number of important issues related to their jobs, including recent fatal encounters between law enforcement and Black citizens."[5] Black officers were about twice as likely as white officers to say that recent deaths of blacks during encounters with police are signs of a broader problem and not sporadic isolated incidents. This study was done with 8,000 sworn officers working in departments of 100 offices or more. On a broader question of overall race relations in American society, the overwhelming majority (92 percent) of white officers say the United States has made the changes needed to give blacks equal rights with whites; just 29 percent of Black officers agree.

A (*New York Times*) nationwide survey in August 2017 found that substantial percentages of blacks see unfair treatment in many aspects of life, especially by the criminal justice system. Fully 70 percent said that blacks in their community were treated less fairly than whites in dealing with the police, and about as many (68 percent) said the same about the courts. By contrast, just 37 percent of whites said blacks were treated less fairly by the police and 27 percent said they received less fair treatment by the courts.[6]

Because of socioeconomic differences, blacks do not receive equality in a number of areas, which might lead to frustration and more frequent coming in contact with law enforcement than whites. The fact is that there is extreme poverty in the United States cities and there is an underground economy that supports many of these neighborhoods: drugs, prostitution, and illegal goods being sold, to name a few. Police have to deal with these realities. This is not just about drugs or law enforcement; poverty plays an essential role in all of this. Jens Ludwig, an economist at the University of Chicago who also directs the Crime Lab there, points out: "Living in a high-poverty neighborhood increases risk of violent-crime involvement, and in the most poor neighborhoods of the country, fully four out of five residents are black or Hispanic."[7]

This being the case, for reasons that may well include police bias—African-Americans have a very large number of encounters with police officers. Every police encounter contains a risk. The officer might be poorly trained, might act with malice, or simply make a mistake, and civilians might do something that is perceived as a threat. The presence of guns exaggerates all these risks. Such risks exist for people of any race—after all, many people killed by police officers are not black. But having more encounters with police officers, even with officers entirely free of racial bias, can create a greater risk of a fatal shooting.

People who live in inner city, poverty-ravaged communities carry a degree of frustration in their day-to-day living; adding racial profiling and "stop and frisk" practices can lead to violent behavior of all parties—the police and the people stopped. Black officers are more empathic to "stop and frisk" because they understand the underlying psychological problems and socioeconomic problems of black citizens, especially black males. Does the ability to stop and frisk go too far? Many police departments are at odds with the public in certain neighborhoods concerning what some people deem unwarranted stops. People in high crime areas and in areas with high minority populations often complain they are stopped and questioned at a disproportionately higher rate than their counterparts in other areas of the city.

It can be argued that it is wrong to blindly accept profiling as simply an effective tool in fighting crime. To do so does not consider the social costs. Throughout our history as Americans, blacks have been treated as second-class citizens. To racially profile only continues this stigma. It deeply polarizes our nation and continues tensions between police officers and the communities they are sworn to protect. It comes at a moral cost to our community and perpetuates overt racism and the notion that those who look different from ourselves are to be feared. This raises the hypothesis that anyone who is different from the majority can be expected to be bullied, subjugated, oppressed, and feared.

Often, some patrol officers complain about an unwritten quota system that rewards officers with promotions based on the number of stop and frisks they perform that uncover drugs or guns. While officers believe the stop and frisk law is a useful crime fighting tool, they also feel the law can be overused in an effort to boost statistics. Moreover, stop and frisks may reduce crime by scaring criminals into thinking they might be stopped at any time, but it also scares law-abiding citizens. This further alienates good citizens and strains the relationship between the police and the community.

In order to understand the broader problem of deadly police encounters among black citizens, you would have to understand, as black officers do, the unique and troubling aspects of growing up black in America, which include black cultural influence, family life, educational systems, being on the lower rung of the economic ladder, and self-concept.

African-American men have faced negative stereotypes and images since the time of slavery when they were used for hard labor and mating purposes. Black men have been judged according to their physique, skin color, intelligence, and language. The effects of images and stereotypes have led to "invisibility syndrome" among black citizens.[8]

African-American men are both feared and ignored by whites, leading them to feel as if others are treating them as if they have no worth. This invisibility syndrome is the inner struggle that one's abilities and personality are not valued. While men are socialized to be providers, they may feel frustrated by sociopolitical constraints because of experiences of racism and oppression.[9] A New York City cab driver offered this wisdom: "You know, when you are afraid, you draw in those things you fear; when you are kind you draw that in as well … so it's best to be brave and kind."

Being stopped by the police for no apparent reason other than "Driving While Black" can cause a high degree of frustration and anger that perpetuates the feeling of invisibility and worthlessness. This can escalate to harsh words and flared tempers. While a black officer might recognize the reaction, he or she is helpless to do anything about a problem that has plagued black men for centuries.

Black officers are much more likely than white or Hispanic officers to say they worry more that officers will not spend enough time diagnosing a situation before acting (61 percent of blacks vs. 37 percent of whites and 44 percent of Hispanics). Overall, blacks and department administrators (59 percent) are the only two major groups in which a majority is more concerned that officers will act too quickly than worry that they will wait too long before responding to a situation.

Saying outright racism likely plays a role in black males being killed at a higher rate than the national average doesn't mean it plays a role every time an officer kills a black person. But other data are unequivocal. Police killings *are a race problem*: African-Americans *are* being killed disproportionately and by a wide margin. And police bias may be responsible. But these data do not prove that biased police officers are more likely to shoot blacks in any given encounter.

If the major problem is then that African-Americans have so many more encounters with police, we must ask why. Of course, with this as well, police prejudice may be playing a role. After all, police officers decide who to stop and frisk and who to profile.

So are the fatal shootings all isolated incidents? Black officers disagree. Black officers tend to lean to the practice of not stopping black drivers unless there is a legitimate reason to do so; they know the potential risk of stopping a black person for no reason other than that they are black and "might" have warrants, drugs, or weapons in their vehicle; if they approach the person in the vehicle who then asks why they are being stopped, then sometimes a harsh exchange occurs between the person being stopped and the police officer. It goes back to "invisibility syndrome," seeing black men as worthless, with no value and no pride.

Researchers say that these and many other factors might underlie the disparity in the killing of black men, but that identifying how much each factor contributes to the burden of police violence borne by black Americans is not possible based on the data available. Part of the broader problem is that black citizens, in general, have many stereotypes to overcome; however, black men have a particularly hard time because of the aggressive persona they have been labeled with. They have been portrayed as lazy, don't like to work, drug dealers, sexual predators, they like white women, and they are a threat to society. The stereotypical concept of the "Mandingo"[10] asserted that "Negroes with all the passions, emotions and ambitions, are subservient to sexual instinct"; so I guess that means they are likely to rape too. These are stereotypes that are still around today and blacks are more likely than whites to be "labeled" a certain type and black police officers know this.

Do white parents warn their children to comply with all questions and orders for police officers, as black parents do? Do white parents teach their children to show unpretentiousness to police officers, as black parents do? Do white parents caution their children to keep their hands on the steering wheel or in sight when stopped by a police officer, as black parents do? Black parents do because they have become accustomed to seeing their children stopped, questioned, beaten, taken to jail, and perhaps shot. White parents don't have the same fear and perhaps don't understand it. To a black parent the fear is real because they witness it in their neighborhoods, on the streets, and in the parks. They have grown to fear the police based on their own experiences and historical facts.

Second, the choice of where to police is mostly not up to individual officers. Police officers tend to be most active in poor neighborhoods, and African-Americans disproportionately live in poverty. In fact, the deeper you look, the more it appears that the race problem revealed by the statistics reflects a larger problem: the structure of our society, our laws and policies. The war on drugs illustrates this kind of racial bias. African-Americans are only slightly more likely to use drugs than are whites. Yet, they are more than twice as likely to be arrested on drug-related charges. One reason for this is that drug sellers are being targeted more heavily than drug users. With fewer job options, low-income African-Americans have been disproportionately represented in the ranks of drug sellers. In addition, the drug laws penalize crack cocaine—a drug more likely to be used by African-Americans—far more harshly than powder cocaine.[11]

A YouTube video, "A Conversation with My Black Son," depicts black parents discussing wisdoms they give to their sons in the event they are stopped by police. Their comments reflect the common feelings and experiences shared by most black parents.

- At some point you *will* be pulled over, it is not a matter of if, it is a matter of when.
- It's frightening, and I'm being very light when I use the word "frightening."
- I've had a few run-ins with police and being completely innocent. So people pull us out of the car, throw us on the floor. It's February in New York. So it's like snow, slush, and stuff on the ground. Put their knees in our back, put guns to our head.
- As I'm putting my hands on the steering wheel so I don't make the police nervous, I realized how nervous I was, and then I realized my children were nervous.
- If something goes wrong, your first line of defense is to go to the police … if you are a Caucasian, unfortunately it can't work for black children.
- He is going to turn into a large, scary black man. That is not who he is, but how he will be perceived.
- Some people say you have to explain racism to him before he experiences racism himself. But when is that?
- It doesn't mean every police officer is inherently a bad person. But what it does mean is that the police force, that institution, does not look out for your best interests.[12]

On September 18, 2016, San Francisco 49ers football quarterback, Colin Kaepernick, in a game against the Carolina Panthers, went to one knee rather than standing during the playing of the National Anthem prior to the beginning of the game. That singular act ignited national controversy that raged across social media and standard news media. People who had never heard of Kaepernick took sides, usually broken along racial or political lines. Many (mostly white conservatives) felt he was offending the American flag and anthem plus showing disrespect to the military, past and present. The other faction (most African-Americans and white liberals) took his kneeling gesture as he meant it—a protest against occurring police violence.

> Kaepernick's social media posts flared with urgent intensity, though, when black men were killed by the police on back-to-back days in early July 2016. […] "This is what lynchings look like in 2016!" Kaepernick wrote on Instagram and Twitter when a video of Alton Sterling's death became public. "Another murder in the streets because the color of a man's skin, at the hands of the people who they say will protect us. When will they be held accountable? Or did he fear for his life as he executed this man?"[13]

In the media, criminal stereotyping is also a problem. TV newscasts show African-Americans being arrested and under arrest, living in slums, being pimps and prostitutes, on welfare and in dire need of help from the community. Violent acts committed by a person of color often show the person as sinister, more so than their counterparts.

In film, black male actors play drug dealers, pimps, con artists, and other forms of criminals. This also fuels the racial stereotypes that black men are dangerous and drawn to illicit activities. Racial and economic injustice makes it more difficult for young black men to evade a prison term, and policies such as "stop and frisk" and racial profiling make black men a target of the authorities. First, the police are at least in part guided by suspect descriptions. And the descriptions provided by victims already show a large racial gap: nearly 30 percent of reported offenders were black. So if the police simply stopped suspects at a rate matching these descriptions, African-Americans would be encountering police at a rate close to both the arrest and the killing rates.

Laws and policies need not explicitly discriminate to effectively discriminate. As French poet Anatole France (1844–1924) wrote a century ago, "In its majestic equality, the law forbids rich and poor alike to sleep under bridges, beg in the streets and steal loaves of bread."[14] We will not sharply reduce police killings of African-Americans unless we understand the social institutions that intimately tie race and crime. In her book, *The New Jim Crow: Mass Incarceration in the Age of Colorblindness*, Michelle Alexander argues that the American criminal justice system itself is an instrument of racial oppression. "Mass incarceration operates as a tightly networked system of laws, policies, customs and institutions that operate collectively to ensure the subordinate status of a group defined largely by race,"[15] she says.

But we all have to be fair and agree that individual police officers did not set these systematic laws and policies that limit opportunities and create the harsh sentencing that turns minor crimes into lifetime sentences. We should strive to eliminate police prejudice because it is wrong and because it undermines our democracy.

Let us not forget that there are also structural problems underwriting these killings. The problem of police killings of unarmed black victims should not be viewed merely as a problem of flawed action on the part of individual police officers, but more as a consequence of the broader problem of structural racism and, as a country, we are all responsible for that.

Racial and religious beliefs are factors in creating societal fear very often manifested by leadership of the government. A striking example is the society created by Hitler, the Nazis and Hitler's dreaded *Schutzstaffel* (SS—literally meaning "protection squadron"). Under anti-Semitism rule, Jews and other non-Aryan peoples suffered fearfully that one day a group of SS would take them into custody to be sent to "the camps."

Nineteen Eighty-Four and "Big Brother"

George Orwell's 1948 novel *Nineteen Eighty-Four* (very reflective of Hitler's Nazi party) reveals a totalitarian society ruled by a government controlling every aspect of the (life) of the citizens who resided there.

> Communication, personal beliefs, and individual loyalty to the government are all controlled by the inner Party in several ways. Thought, speech and actions are all monitored by political officials in one way or another.[16]

In the novel there is an on-going, never-ending war, and citizens voluntarily give up freedom in the name of security provided by the government. Though fearful of the far away enemy, they are more fearful of the government. The vast majority, by default, are supportive of the government, even though they fear it. Dissidents are not sent to a concentration camp but to "Room 101," where they are confronted with the one thing they fear the most until they are "cured" and believe that they "love Big Brother"—the government.

> Newspeak [political correctness] is a modified version of language that is enforced upon the people in order to limit their expression ... "It's a beautiful thing, the destruction of words ... You haven't a real appreciation for Newspeak ... Don't you see that the whole aim of Newspeak is to narrow the range of thought" (Orwell, 46).[17]
>
> The Thought Police are a group of devout followers that spy on people, and determine whether or not they are thinking negative thoughts or have a desire to overthrow the Party. If so, they are accused of thought crime, and the punishment is death, either by hanging (a public display, used to keep the general population afraid, and in line) or parting with your life and being in a dark, dank room (Room 101) while being tortured by a Thought Police official. In either case, the outcome is not a positive one, and the Party's effectiveness is rather surprising. Nary an individual dare to commit thought crime.
>
> The Party also used electronic devices to keep tabs on the inhabitants of their kingdom. Similar to a typical television, the "telescreens", as they were called were like cameras, watching every move made by any one at any time. It was uncertain whether party officials could watch everyone, at every time, but out of fright, citizens assumed they could ...the assumption was generally made that "yes, you are being watched."[18]

Thus the term "Big Brother!"

"Big Brother" was the name of the faceless government that controlled the people in Orwell's dystopian world. In today's society it is very common to hear the expression, "Big Brother is watching you." A surveillance society has been created with the instillation of cameras and monitors as a matter of routine in every place, ranging from the baby's crib to every street and household in a city as a result of the fear of crime, violence, and terrorism; and in many circles, the government and the police.

The culture of the people populating that fictitious realm of *Nineteen Eighty-Four* is loveless, gray, bleak, uniform, and non-religious, their total beliefs, values, and practices formulated and enforced by the ruling party. Rules and orders are blindly followed out of fear and "brainwashing." Procreation is forbidden without consent of the Party. The children who are born and raised in that environment do not initially experience fear; blindly accepting all dictates of "Big Brother" as a normal lifestyle; believing, without question, in the security, programing, and standardized uniformity afforded by the government.

> Fortunately, a fine line exists between the free, democratic society under which we live, and the closed-minded, totalitarian regime that fictionally existed here. Today's people are much more intelligent, free-minded, and aware of their surroundings to fall victim to such a terrible, controlling trap [or are we?]. We like to maintain that assumption, anyway; we must believe it to continue with our daily activities and lives. Fortunately for us, that's the way it is, and Orwell's novel was just fiction, which we can put down at any time, look around and feel assured that we are free and safe. Unless, we really are being watched?[19]

The terms "political correctness" and "surveillance society" found a voice in the later twentieth and early twenty-first centuries. The intent is to eliminate words from the language that may have a tendency to offend others. The corrective stance is because of fear of offending, thus causing hostilities and division in society. The consequence of the new wording has become very favorable in all circles of society and politics but has also had a dampening effect on the freedom of word choice (freedom of speech). A major consequence has been the "hyphenation" of native group identification such as *Italian-American* and *African-American*, though a person was born, raised, and spent his entire life in the United States as were his parents, grandparents, and others dating back several generations.

A Politically Correct Surveillance Society

Does surveillance equal security? Yes, we know that the government has agencies and systems monitoring telephones, emails, and all social media searching for terrorists and criminals. The frightening thing is that the net grows larger and more widespread. Yet while we can assume there is no government conspiracy, we also must be cognizant that they know more about us than we realize. Data are collected from thousands of sources and entered into databases, all very similar to Orwell's *Nineteen Eighty-Four*. Combine all the data collection with cameras equipped with audio, zoom, tilt, and pan plus facial recognition everywhere, and it can be readily seen that contemporary society has a fearful personality.

Who is "Big Brother"? He is us. Today's world is such that nearly everyone, even children who still may be in preschool, has a cell phone that is always in a pocket or in a hand. In every telephone (and computer) is a high resolution camera available to record, in still photos or moving video with sound, every instance that draws their attention and interest. No one knows when they are being videoed or photographed. It must be assumed, no matter where or under what circumstance, a camera has its sights set on the scene. Everyday citizens take photos of someone being bullied and beaten but are so intent on making the recording, they fail to come to the assistance of the victim. Could that failure be attributed to fear of becoming involved, possibly being injured or saddled with a law suit or possible arrest?

A culture of voyeurism has now grown a generation of adults who have been exposed to being monitored on a 24-hour-a-day basis from their earliest moments of birth so that a second thought about the loss of privacy and movement toward Orwell's prediction never enters their mind. They have come to believe cameras and monitors are a normal fact of life. They have become impervious to the potential dangers hidden behind every lens and have no fear that monitoring is merely nothing more than a security blanket. In a study, students in several college classrooms were asked about their thoughts regarding cameras placed in public areas. In nearly every reply came the question, "What is there to be afraid of if you are not doing anything wrong? The cameras are to help keep us safe."

Where It All Began

There is a small city in southwestern New York State (Olean, estimated 2017 population 13,711[20]) which, in September of 1968, became the first in the country to install video cameras along its main business street in an effort to fight crime. The use of closed-circuit TV cameras piping images into the Olean Police Department propelled Olean to the forefront of crime-fighting technology and societal surveillance. In March, 2008, a reporter, John Eberth for the local newspaper, wrote a human interest story about the installation of security cameras on the main street of the small town. His column is quoted:

Within a year of the cameras being installed on North Union Street, more than 160 police chiefs from around the country visited Olean to inspect the system as well as a colonel from the U.S. Army Weapons Division; Dr. Donald E.J. MacNamara, a professor of criminology at John Jay College of Criminal Justice; representatives of the Israeli Defense Force; and Maj. Gen. Somjai Jenyavanija, head of the national police force of Thailand.

Mike Arnold is a retired communications industry executive now living in California. In 1967 he was the president of Allband Cablevision Co. bringing then-new cable television technology to the city of Olean. His company installed the cameras on North Union Street in the downtown business district to look for crime.

Mr. Arnold said the level of interest in the system took him by surprise—he was just trying to help downtown merchants protect their stores. "I mean, we had people coming from Israel and Thailand to see how the system worked," he said chuckling. "The New York Times came here twice. They thought people would rise up and scream about it because of what they called 'Big Brother syndrome.' But the reporter was here for two weeks and he couldn't find anyone who was against it."

Mr. Arnold said it all started innocently enough. He was running the cable television system in Olean at the time. During an Olean Exchange Club meeting, he listened with interest as the Olean SEARS store manager complained the company had ordered him to buy a guard dog for the store that he would have to care for. SEARS in Olean had recently been broken into and company officials in Chicago didn't want to be ripped off again.

Mr. Arnold suggested putting cameras on utility poles downtown as a deterrent to crime instead. The idea was a hit with the other business owners at the meeting, but then Mr. Arnold began wondering what it would cost. In the end it took more than 1.4 million dollars over two years in 1968 dollars to design, install and run the system.

Mr. Arnold said it would never have happened without the OK of Alfred Schwab, owner of Allband's parent company in New York City, the Television Communication Corp. of America. Mr. Arnold said when he discussed his idea with Mr. Schwab, he was completely in favor. "When I told him what we wanted to do, Al Schwab backed us 100 percent. He told us to make it work no matter what the cost." Mr. Arnold said. "All the funds for this came from the company. It was a different era back then."

Mr. Arnold worked with company engineers to build the cameras and design the system. It was among the first efforts outside of the military to build a durable, all-weather video camera system. The cameras ran 24 hours a day, feeding images into a bank of televisions at the Olean Police Department. The cameras panned left and right on a timer, overlapping their coverage of the street.

Mr. Arnold said. "At the time, New York City had one camera and it was stationary; it couldn't pan. In reality, Olean is the birthplace of television surveillance in the country."

Mr. Arnold said the camera system did suppress crime on North Union Street. "During the time the cameras were up, there were no break-ins on North Union Street," he said. "All the crime moved to West State(Street)."

The cameras were up and running for 18 months on North Union Street. The system died with the demise of Allband's cable franchise in Olean. The system had 900 customers in Olean, operating from 1964 until the early 1970's.

"It was a three-channel system to start out with," Mr. Arnold said. "Then they came out with 36 channels including WPIX out of New York and everybody was saying, 'what are we going to do with all these channels?' We charged $5.50 per month, which people thought was excessive."

Eventually the franchise was bought out by Warner-Amex Cable Co., which later merged with Time to become Time Warner Cable. Mr. Arnold is proud he had a hand in designing the camera surveillance system. "I was 26 at the time and just felt lucky as heck to be a part of this thing," he said. "I think it was just a little bit ahead of its time."[21]

Summary

The contemporary world is a fearful, surveillance society manifested by crime, racial discordance, bigotry, distrust, and authoritarianism. The greatest social divide is the long-running issue of stereotypes, misunderstanding, and ignorance. They become a festering sore in police–minority relations of distrust, fear, violence, and often unjustified shootings. Unfortunately, cloaks of aspersions are thrown over the entire field of policing when a member of a minority group is beaten or killed by an officer who rightly or wrongly believes his life is in immediate danger.

Complicating society is the growing fear of minorities by white men and women who feel intimidated by the presence of a person of color, especially Hispanic immigrants and African-Americans. In 2018 the growing mistrust between whites and others has resulted in unnecessary and harassment-type calls to the police from people who feel threated by the mere presence of a person of color.

There is a growing sense of fear of crime and terrorism in the country resulting in a spreading monitoring of people in public areas, similar to the dystopian world of George Orwell's novel, *Nineteen Eighty-Four.*

Review Questions

1. How do subtle prejudice and ignorance result in over-reaction between police and persons of color?
2. Do you believe white parents should find it necessary to give their children the same advice about interacting with police as black parents do?
3. Do you believe Colin Kaepernick "taking a knee" was unpatriotic or in a just cause?
4. What explains why more black males are arrested and/or shot than are white males?
5. Do stereotypes play a role in creating an atmosphere of fear? Explain.
6. Does a society of surveillance, including police wearing cameras, signal a more secure society?

Notes

1 http://mpdc.dc.gov/release/overcoming-fear-building-partnerships-toward-new-paradigm-police-community-race-relations. Friday, December 11, 1998; New Jersey Law Enforcement Summit; Chief Charles H. Ramsey; Metropolitan Police Department, Washington, DC.
2 Retrieved July 28, 2018, www.dictionary.com/browse/microaggression.
3 Lovgren, Stefan. "Race Affects How We Learn to Fear Others, Study Says,' National Geographic News July 28, 2005; Retrieved March 21, 2018.
4 www.cbsnews.com/news/smith-employee-who-called-police-said-black-student-looked-out-of-place/. Retrieved August 23, 2018.
5 Gramlich, John. "Black and White Officers See Many Key Aspects of Policing Differently" (2017), Pew Research Fact Tank: www.pewresearch.org/fact-tank/2017/01/12/black-and-white-officers-see-many-key-aspects-of-policing-differently/. Retrieved October 14, 2019.

6 Carbado, Devon, Kimberly Crenshaw, and Cheryl Harris (August 14, 2013), "Racial Profiling Lives On." *The New York Times* p. 23A: www.nytimes.com/2013/08/15/opinion/racial-profiling-lives-on.html. Retrieved October 14, 2019
7 Mullainathan, Sendhil; "Police Killings of Blacks: Here is What the Data Say." *The New York Times*; October 16, 2015.
8 Franklin, Anderson J. *From Brotherhood to Manhood: How Black Men Rescue their Relationships and Dreams from the Invisibility Syndrome*. New York: Wiley (2004).
9 Neville, Helen, B. Tynes, and Sharon Utsey. (2009) *The Handbook of African American Psychology Intimate Relationships of African Americans*, p. 118. Sage: Los Angeles, London (2013).
10 Mandingo is defined as people who live along the upper Niger River Valley in West Africa and to languages that are spoken in this region (YourDictionary.com). In terms of this paragraph, *Mandingo* is a 1975 film, based on the novel *Mandingo* by Kyle Onstott and upon the play based thereon by Jack Kirkland. "Mandingo is a man of native origin. Specifically known for the size of his penis. His penis is big, black and has plenty of girth: Mandingo is a man with many sexual attributes." www.answers.com/Q/What_is_Mandingo. Retrieved December 26, 2018.
11 Op. cit. Carbado, Devon, Kimberly Crenshaw, and Cheryl Harris (August 14, 2013), "Racial Profiling Lives On." *The New York Times*. p. 23A.
12 Retrieved August 4, 2018 from *The New York Times*; "A Conversation With My Black Son," Op-Docs, *The New York Times*. Published March 17, 2015.
13 Branch, John; "The Awakening of Colin Kaepernick"; September 7, 2017 www.nytimes.com/2017/09/07/sports/colin-kaepernick-nfl-protests.html. Retrieved August 29, 2018.
14 France, Anatole; *The Red Lily* (1894), ch. 7.
15 Alexander, Michelle; *The New Jim Crow: Mass Incarceration in the Age of Colorblindness* (2010); The New Press.
16 *The Traits of Society in 1984*; www.123helpme.com/assets/15559.html. Retrieved September 12, 2016.
17 Ibid.
18 Ibid.
19 Ibid.
20 www.worldpopulationreview.com/us-cities/olean-ny-population.
21 Eberth, John; "Nation's first surveillance cameras were installed in Olean." *Olean Times Herald*, October 8, 2008.

7

LIVING IN A WORLD OF FEAR

> We are living in a world of fear. The life of man today is corroded and made bitter by fear—fear of the future, fear of the hydrogen bomb, fear of ideologies. Perhaps this fear is a greater danger than the danger itself, because it is fear which drives men to act foolishly to act thoughtlessly, to act dangerously.[1]

Schools

> The earliest known United States shooting to happen on school property was the Pontiac's Rebellion school massacre on July 26, 1764, where four Lenape American Indians entered the schoolhouse near present-day Greencastle, Pennsylvania, shot and killed schoolmaster Enoch Brown, and killed nine or ten children (reports vary). Only two children survived.[2]

From January 10 to October 29, 2018, 36 school shootings in the United States happened, with 43 fatalities and 81 others injured.[3]

Fear in a community is often bred from news media stories and baseless rumors, becoming hysterical witch hunts that do more damage than the actual incident. Perhaps the primary example is unfounded, generated by half-truths and innuendo. It is becoming more and more normal to see urban schools surrounded by 7-foot high fences with the only means of entry and exit being through a police or security officer monitored gate. Many school districts have their own police departments and assign officers to patrol the hallways while school is in session. In some areas, the schools appear more like a prison than an institute of learning. Schools are also considering training and arming teachers or hiring armed former military personnel to patrol the hallways and grounds of the school.

A very rural school district in a small town with a population of slightly over 1,100 people recently initiated an identification system intended to identify and exclude any "registered sex offenders" who may attempt to enter the school building. The intent is commendable but in actuality there are some points not considered that will have unintended consequences. For example, the number of sex offenders entering a school looking for victims is so miniscule that the percentage of a chance of that occurring is so low the number wouldn't even register on any scale. Over 90 percent of sexual molestations of children come from within the family—a step father, a brother, a cousin, an uncle, a father, or perhaps a neighbor. Living in such a small community, a registered sex offender may have a child or children in the school, but his sexual registration requirement and offense is a well-kept secret. Not only will his secret become public, but he will be barred from attending any school functions his child is participating in. There is also the possibility the reason he is on the list was such a long time ago that even his family doesn't know about it. Often a person is placed on the must register sex offender database for something that happened several years prior, perhaps he was 20 and impregnated a 17-year-old girl.

According to the supervisor of the school district:

> [The school district] is pleased to announce that we will begin using the XXXXX Visitor Management System in all of our district schools to build on the district's program of campus safety for students and faculty. Part of keeping students and faculty safe is knowing who is in our building at all times and the system will allow us to do that. The system will better allow us to track visitors, contractors, and volunteers in our schools and provide us with a safer environment for our students and staff.
>
> Upon entering a district building, visitors will be asked to present a valid state-issued ID, which will be scanned into the system. The XXXX system will check to ensure that registered sexual offenders are not entering our buildings. It is important to note that the system only scans the visitor's name, date of birth and photo for comparison with a national database of registered sex offenders. Additional visitor data from the driver's license is not gathered nor is the system connected to any other system such as the Department of Motor Vehicles. Therefore, any other information on the ID is not scanned by the system and is not accessible to any of the users. Once entry is approved, (the system) will issue a badge that identifies the visitor, the date, and the purpose of his/her visit.
>
> A visitor's badge will not be necessary for those who visit our schools simply to drop off an item in the office or pick up paperwork. In the event that a person does not have identification, he/she can still be given access to the building, but will be escorted by a school staff member.
>
> The safety of our students is our highest priority and the visitor management system provides a consistent way to aid in keeping away people who may present a danger to our students. Therefore, all visitors to the school must be scanned into the system and obtain a visitor badge. Thank you, in advance, for your understanding and support of enhancing school safety protocols in our district.[4]

The administration moved its offices to the ground floor near the locked entrance. Visitors must be identified (cameras and monitors) plus present a photo identification and explain the reason for the school visit prior to being "buzzed in." Then they must stop at the administration office to sign in. Some schools have gone so far as to install "Sally Ports." A sally port is generally defined as a secure, controlled entryway to a fortification or prison. It normally consists of entry through one locked door into a small waiting space until a second door is unlocked.

> Our children have been exposed to fencing—some times multiple layers of fencing—channelizing them through a singular pedestrian entrance, metal detectors, mono-directional, full height turnstiles, followed by bag searches, issuance and requirement to display an ID at all times, unarmed and armed security, and it doesn't seem to have negatively impacted them, nor diminished their desire to attend the locations where such security measures are absolutely commonplace—which include amusement parks, sports stadiums and arenas, as

> well as entertainment venues. While having witnessed first hand the evolution from first implementation to the current proliferation of such measures at all manner of facilities and venues frequented by kids of various age groups, I remain unaware of a singular instance where a parent has denied their child the privilege of attending any of the venues and events where such measures exist based on a belief that those security precautions would have a negative impact on their child's experience or psyche, nor have I heard the comparison made to a prison. Likewise, I am unaware of any parents protesting the implementation of such measures at those venues and events.[5]

Unsaid is that no one ever vetoes a security measure believing that should even one incident occur or be deterred over many years of heightened security, the expense and inconvenience would be worth all the restrictions. Unrealized are the unintended consequences of raising generations of people who embrace every security matter in lieu of practicality and loss of freedom of movement. Is it possible that future generations will be comfortable only if they are surrounded by walls, fences, gates, procedures, and monitoring? The result will be that "authority" will control movement of everyone. That sounds pretty much like institutionalization and prison, perhaps a dystopian future.

Metal Detectors in School

"Metal detectors coming soon to all Duval County high schools."[6] This is the headline in a local newspaper in Florida with the school district putting its faith in even more technology to "ensure the safety of students and staff."

> All Duval County high schools will have walk-through metal detectors and hand "wands" for checking students for weapons in the coming months, district officials said Monday night ... The decision to buy metal detectors appears to reverse a long-held position held by Duval County Schools' previous leaders, including former Superintendent Nikolai Vitti, who were reluctant to take that measure in part because of cost and a belief that relationships between students and school staff were better options for unearthing weapons in school.
>
> Director of the district's police force Micheal Edwards said ... that since the mass shooting at a Parkland, Florida high school and more recent incidents in Jacksonville, including shootings involving Duval students and former students, things have changed. Superintendent Diana Greene said each high school, though it has many entrances, will receive two walk-through metal detectors but it's not yet clear where they will go. "We'll need the school to tell us how the traffic flow works and where students enter the building," she said.[7]

The counter point to all these drastic security measures is that they have now become commonplace and people no longer take note of the growing push to surround ourselves with them. A country boy and an old western cowboy song would say: "Don't fence me in, Can't look at hobbles and I can't stand fences, Don't fence me in."[8]

Where does it end? No one will ever vote against any type of security measure. Whatever someone suggests, it will be implemented. Security is limited by only two things, imagination and finances. If it can be imagined and paid for it will be done. Every time a security measure is adopted it means the security noose is closing and another limitation on free movement is imposed. Shouldn't we be fearful of all the security measures?

Coddled Kids

> In recent years, there has been a growing awareness of the phenomenon of students in higher education needing protection from ideas that can upset them: trigger warnings, safe zones, limitations on free speech, and protection from microaggressions. Students' belief in their right to be protected from feeling offended has even led some to resort to physical violence to prevent speakers from appearing on campus.[9]

> Many college students are showing themselves incapable of accessing even basic internal coping skills. As a result, today's colleges and universities are becoming equal parts psychologists, in absentia parents, and even academic scapegoats (when students don't get the grades they thought they would, for example).
>
> "The idea of fragility is now an overarching theme in kids," said Lenore Skenazy, founder of FreeRangeKids.com. "This comes from a society that makes adults fear everything, and constantly tells them they are not doing enough and not worrying about enough, so they parent their kids accordingly."
>
> In the continuing infantilizing of America's young people, colleges report having to warn kids when they are going to talk about something that might be mentally troubling in class, with a so-called "trigger warning." That way, in theory, a past victim of abuse or violence would be warned when graphic or violent material might be coming their way [the students who would be affected are allowed to leave the classroom].
>
> This isn't that surprising, considering what is regularly happening: Students are swarming their campus counseling offices for once-trivial life occurrences, such as being made to feel bad by another student. In one "funny-if-it-weren't-so-sad" case, two students saw a mouse, dialed 911, and then sought counseling for the resulting trauma—an apparent case of rodent PTSD. [...] In the stampede to campus counseling offices, student also complain of anxiety over multiple issues.
>
> "The mental health common cold for this generation of students is anxiety disorder, whereas for previous generations it was depression," Kip Alishio, director of student counseling at Miami University, told USA Today. [...] What is happening to the youth of America? Why are they turning to Jell-O when faced with the slightest bump in the road?
>
> "Students haven't developed the skills to soothe themselves, because their parents have solved all their problems and removed obstacles. They don't seem to have as much grit as previous generations," Dan Jones, a past president for the Association for University and College Counseling Center Directors, wrote in The Chronicle of Higher Education. [...] Another troubling emergence in the psychology of today's young adults is their fear of learning the lessons that failure offers. Failure, for young America and their parents, is the new taboo.[10]

Parks and Recreation

Fear for the lives, security, and safety of their children has prompted activist parents to seek injury-free playgrounds. Areas beneath swings, climbing bars, and other equipment are padded with rubberized, shock-absorbing material or layers of sand to minimize impact should a child fall. In lower level baseball leagues (little league to high school) and football, fields with natural turf have been turned to shock-lowering artificial surfaces. Laws have been written to prohibit sex offenders from living within 2,000 yards of public parks and schools. A Los Angeles councilman proposed a "Law (that) would ban lone adults from playgrounds." According to the councilman, "The law would ensure that the city's playgrounds are a 'safe haven' for kids."[11] As proposed, the law would prohibit an adult not accompanied by a child from the playground area. Currently the areas are separated from the "adult" area of the parks by fences and gates, and marked with sand and rubber mats. According to many, such laws are unnecessary and lead to more evidence of the formation of a "nanny" state.

Homes

The safest place in a person's life is assumed to be his or her home. Tranquility, green lawns, congenial neighbors are historically the stereotype of the "American dream." Of course dreams are not reality. As far back as the 1960s a few residents in high crime-rate areas in cities placed steel doors and window screens in their houses. It was widely reputed that these were "drug houses," but whether this was true or not the result was they afforded the home owner a sense of safety.

As we move deeper toward the third decade of the twenty-first century, the "fortification" of homes has increased a hundred fold. Planned communities with gated guarded entrances, private surveillance cameras and alarm systems in the houses, private patrol operators doing routine patrol checks with armed response. The marketplace for apartments has grown substantially with developers and marketers using the concept of secure buildings as an added benefit. The trickle has grown to full-fledged flooding of security-related technology built into newer apartments and office buildings with promoted sales being dependent as much on marketing security from fear as other attributes and benefits.

To enter newer apartment complexes has become extremely difficult. Parking lots and garages are dotted with "speed bumps" to discourage driving over 5 miles per hour to thwart any speedy "get aways." There is no more simply walking to an apartment and knocking on the door; the front doors are locked and latched with controlled entry boxes with specific codes for each apartment. The locks are doubly reinforced with steel plates over the latches. Once entry is gained, there is a security officer monitoring the premises with closed circuit cameras and logging in all visitors and controlling the elevators. Very often there are private security officers patrolling the grounds and hallways.

The people renting and buying the new security "wrapped" buildings are only sublimely aware of the phantom fear they are displacing with all the security they are paying for. When they arrive back home they double lock the doors and turn on their alarm system. Fear may not enter their mind but the psychology of feeling secure gives them confidence in their wellbeing.

Workplace Fear

A new employee reports for her first day of work. It doesn't matter the setting; it could be a business office building, an educational institution, a factory, a retail store, but the first thing that happens is she reports to have her photo and identification taken and sealed in a laminated "identification card" on a lanyard she must constantly wear around her neck. Naturally the assumed purpose is for identification of everyone who has authority to be in that environment. Of course the card is also magnetized and encoded with all the required information to allow the cardholder access to the building and specific areas; her movement can be monitored as she enters and leaves an area. It is also so that building or office security can look at the badge and authorize the person to be in the building.

That is a very sound intention, however, casual observation has shown that the badges are very seldom checked by security or co-workers questioning the wearer; especially if the officer gets to know the person and allows entrance simply by facial recognition. As long as the person is wearing something that looks official, no questions are asked. Whatever the intention, fear of an unauthorized person being present has resulted in wholesale wearing of identification cards to the extent that they have become an accepted way of life.

In many instances when a person's employment is terminated the identification card is not returned to the employer, thus allowing potential unauthorized entry. The reasoning for issuing visual identification cards is the fear of violence. In workplace violence cases, an identification card has not stopped one person intent upon hurting others.

To enter government office buildings a person must empty all his pockets, remove his belt, open his briefcase, then walk through a metal detector and be hand scanned with a wand-holding security officer. On passing these tests, the office visitor then must register at a desk and show identification. He is monitored by camera surveillance until reaching his final office destination. There he must be "buzzed" in, then go to a counter with a receptionist behind bullet-proof glass. Banks and other "public" facilities have the same type of glass bullet proofing. Security surrounds us, no matter where we go!

Fear—Living in the Street

> "To survive on the streets, you must live by a whole different set of rules. I survived ten years living on the street pretending I was mentally deranged."
>
> *Homeless female street survivor*

> There are approximately 70,000 homeless persons sleeping on the streets on any given night in Los Angeles, more than any other metropolitan area in the United States—up to three-quarters of them suffering from mental illness or physical disabilities manage to protect themselves from the elements and from predators and tuck into bed, often in plain sight.[12]

Surviving the streets is not easy. Fear of injury, and danger of rape, physical violence, robbery, poor health, and arrest, or harassment by the police, is the existence of those who for some reason are homeless, living in the hostile environment of city streets. There is no law—the only rule is the law of the street—survive by any means possible. Recognize there is evil out there and acts of violence are apt to happen at any moment.

To survive means living by "the code of the street"; street savvy includes being a good criminal, petty thief, etc. Don't talk to police, don't be a snitch even if you see crimes committed; follow a code of silence—"streets talk and people walk"—word gets around and people take action. Surviving means learning to overcome fear and adapting to the unspoken rules, beginning with recognizing other street people—the drug dealers, prostitutes, homeless (mental or not) harmful or passive; learning "good" and "bad" areas of town—divided ethnicities, gangs, political disorganization, and times of day when an area is good or bad (environmental). Be aware of scavengers—they have their own "territory for recyclables," territorial rights of sleeping and camping, and most carry improvised weapons! Street people must not be fearful or show fear as they are living by their wits and courage.

Summary

Fear surrounds us daily in every walk of life. "Fear deprives people of choice. Fear shrinks the world into isolated, defensive enclaves. Fear spirals out of control. Fear makes everyday life seem clouded over with danger."[13] To help alleviate fear, the "hue and cry" is to boost security. Babies are introduced to fear and security from the moment they are born. They are placed in camera-monitored infant beds in the hospital, then in cradles at home monitored by worried parents by intercom and cameras. The rest of their life will be security watched with no thought to personal privacy, believing monitoring and all the precautions are merely normal and natural.

Fear has turned playgrounds and schools into reflections of prisons with fences, monitoring, procedures, and practices. Public buildings and homes have the latest technology to detect and deter any person from committing a "horrific" act. Every public event, from sports events to celebration parades, poses security problems mitigated by a tightened security presence.

Fear has entered and changed the meanings of words so that many have been filtered out of vocabularies to not be offensive or sound threatening. Fear has life fenced in.

Review Questions

1. Have school districts overreacted to potential danger by instituting prison-like security measures to ensure the safety of students?
2. Has the mentality of society gone beyond reason to incorporate security in every measure of living?
3. Do modern parents tend to overprotect and coddle their children?
4. Is the current generation too absorbed in fear and security that the "fiber of youth" has resulted in atrophy of daring and adventure?
5. Has fear and obsession with security deprived people of freedom and choice?
6. Has a "fearful society" been created by a generation of parents who coddle and attempt to ensure the safety of their children? Are today's children protected too much?
7. We teach our children to "never talk to strangers" because of the potential for child sexual abuse. Is this solid and adequate advice? Does it create fear and anxiety in children?
8. How have recent shootings (mass killings) shaped and influenced the United States' attitude toward violence?

Notes

1 www.quotetab.com/quote/by-sukarno/we-are-living-in-a-world-of-fear-the-life-of-man-today-is-corroded-and-made-bitt#OvrMq4HkQLua6xy1.99. Retrieved January 21, 2018.
2 "History of School Shootings in the United States"; www.k12academics.com/school-shootings/history-school-shootings-united-states; Retrieved December 2, 2018.
3 "List of School Shootings in the United States." This article lists in chronology and provides additional details of incidents in which a firearm was discharged at a school infrastructure or campus in the United States, including incidents of shootings on a school bus. This list contains school shooting incidents that occurred on the campuses of K-12 public schools and private schools as well as colleges and universities. https://en.wikipedia.org/wiki/List_of_school_shootings_in_the_United_States#2015_to_present; Retrieved December 2, 2018.
4 School in quotation not noted for privacy reasons but a similar letter from a different school is available here: www.milfordschools.org/pattison-elementary-news/entry/new-visitor-security-system-237/for/. Retrieved October 21, 2019.
5 A Security Consultant and Trainer; Facebook March 24, 2018.
6 Amos, Denise, *The Florida Times-Union*; October 16, 2018.
7 Ibid.
8 "Don't Fence Me In" is a popular American song written in 1934, with music by Cole Porter and lyrics by Robert Fletcher and Cole Porter. It has been recorded by several western music singers and is considered as one of the top 100 western songs of all time.
9 Kalman, Izzy; "The Promotion of Resilience and the Demise of Anti-Bullyism"; *Psychology Today*; January 18, 2018; www.psychologytoday.com/gb/blog/resilience-bullying/201801/the-promotion-resilience-and-the-demise-anti-bullyism. Retrieved January 23, 2018.
10 www.foxnews.com/us/2015/10/15/coddled-kids-crumble-colleges-see-big-lack-resilience-among-students.html?ref=yfp.*LifeZett* October 15, 2015. Retrieved March 22, 2017.
11 Smith, Dakota; "Law Would Ban Lone Adults from Playgrounds"; *Los Angeles Times*; January 2, 2017, p. B1.
12 Garrison, Jessica; "Finding L.A.'s Hidden Homeless"; Los Angeles Times; August 23, 2008, p. A-1.
13 Deepak Chopra, www.behappy101.com/recession-chopra-article.html. Retrieved March 5, 2019.

8

TERROR AND FEAR OF THE UNKNOWN

Paralyzing anxiety is that element whereby potential victims allow their actions to be influenced by fear. One person can wage a reign of anxious fear based on the randomness of the threat!
Fear causes the wolf to look bigger than he is.

German proverb

Witness the terror wrought by the "Unabomber," Theodore Kaczynski. It was not the fact that he killed 3 and wounded 29 others that caused so much fear and concern in the years during 1975–1995. The fear was a product growing from the Unknown. "The oldest and strongest emotion of mankind is fear, and the oldest and strongest kind of fear is fear of the unknown."[1]

The unknown sender of anthrax letters in 2001 generated a similar fear. Prior to September 11, 2001, anthrax was a little-known farm animal disease. Immediately after the terror aircraft attacks of the World Trade Center and the Pentagon, anthrax-filled letters began arriving at the offices of two U.S. senators and several news organizations. The United States began to ask, "Who is next?" and special precautions for mail were taken in government offices and news media organizations. Today, nearly everyone is not only familiar with the term "anthrax" but it has become *le mot du jour* (the word of the day).

The bio-terrorist spreading the anthrax spores might have been a single lunatic fringe criminal, or perhaps part of a terrorist conspiracy, and the number of infected victims very small but that is secondary to the results of spreading fear from an unknown source. Ten years later the FBI identified a potential suspect, but he was never arrested because he committed suicide before the FBI could question him.

Observe the fear of the unknown and unpredictable generated by the "D.C. Area Sniper." In September and October 2002, John Allen Muhammad and Lee Boyd Malvo conducted a shooting spree in the Washington D.C. region that left ten dead and three wounded. The duo paralyzed the entire east coast region for more than three weeks before being apprehended at a truck rest stop in Myersville, Maryland.[2]

The questions on everyone's mind were: "Who is the next victim?"; "When will he strike again?"; Where will he strike?"; "Who is he and why is he doing this?"; "Isn't anyone safe?" The fear became so great that travelers began to avoid the entire Washington D.C. area whenever possible. There was no respite from the daily growing fear. Shooting from the closed trunk of a car, through a small hole in the side of the vehicle, the snipers could strike suddenly from nearly any location and drive away unnoticed and unobserved.

These are valid questions asked about any type of unknown source of potential killers. They could be asked about a serial killer or a rapist. The commonality is the fear generated by the seemingly random actions of an unknown entity. The fear factor of danger lies in its seemingly unpredictable nature, and *randomness of the violence.*

South Los Angeles experienced a community-wide fear when rumors of gangland violence and warfare spread on social and news media, which predicted 100 gang members to be killed in 100 days. The fear quickly escalated from street gang members to local residents. "People were thinking that it was just out of control. Wild, Wild West, that people were dying on a daily basis and that really wasn't the case."[3]

The police responded by saturating the area with patrol cars and providing assurances to the community that the fears were unfounded and based solely on rumors spread by social media. According to police, only one gang-related homicide (one too many) occurred during the week of the fears.

Fear in his victims makes a serial killer or rapist passionate and ecstatic. To spread the most fear and shock, victims must feel that nothing or no one is immune. Spreading fear is the goal of a successful terrorist, even if he is killed in the action! According to the philosophy of the late Chinese Chairman Mao Zedong, "Kill one, frighten thousands."

Potential terrorist targets do not necessarily have to be of strategic importance; they may be selected more for the shock value that can be generated. For example, national landmarks, traditional treasures, museums, military or veterans' hospitals, etc. make excellent points of reference for a terrorist attack. As Kaczynski and the D.C. Snipers spread fear by selecting their victims seemingly at random, the same can be said for terrorists.

Since September 11, 2001, fear is a constant reminder that terrorism is successful, even if the perpetrators of the act are killed. The belief, "It could never happen to me" has been translated to, "It could easily have been me" or "Will I be next?" as we now realize that terror can, and will, strike close to home.

The key words in any definition of terrorism are *violence* and *fear.* Violence and Fear will always be the methods of choice for those who take it upon themselves to influence and/or destroy institutions with which they disagree. Osama Bin Laden, and others, like the current religiously violent groups, Boko Haram and the Islamic State (ISIS), proved that a small, dedicated group of fanatics can bring towering nations to bear and the world to shudder.

Their goals are to spread as much fear as far as possible. They are willing to commit "kamikaze" self-destruction type attacks, such as becoming human bombs or carriers of infectious diseases, dying in the course of trying to destroy government infrastructure or military targets or paralyze a frightened civilian population.

Spreading fear and anxiety makes the victimized audience feel impotent, forcing them to change their ways, surrendering essential liberties like the freedom of movement, in the name of security. Fear is the greatest weapon. One of the best ways to conquer fear is to learn it, understand it, and face it. The best defense against fear is refusing to have your fears influence your actions.

There are many World War II photos and films of soldiers in landing craft approaching the French Normandy beach beginning the Allied invasion of Europe on D-Day June 6, 1944.

German machine gun fire was heavy and thick, with bullets pinging off the landing craft. Each soldier heard that sound, like rain beating on a tin roof, and knew that as soon as the

FIGURE 8.1 U.S. assault troops in an LCVP landing craft approach Omaha Beach, June 6, 1944
Source: www.wikiwand.com/en/Normandy_landings. SC-320901 Photograph from the Army Signal Corps Collection in the U.S. National Archives.

gates of the craft opened he would be nakedly exposed to the withering hell of gun fire. Over 9,000 Allied troops died that day, and over 150,000 died during the entire invasion on subsequent days. But each man, overcoming his fear, bravely stepped into the water and ran toward the beach. Both fear and death were present and the odds of becoming a casualty were very high, yet, not one soldier refused to step into the water and face the bullets coming his way.

In being full of fear and afraid, what is there to be afraid of? The obvious, but wrong, answer is death. When you examine death most people fear it because of the uncertainty of how one will die. Will it be instantaneous or dragged out? Will it come like the gentleness of falling asleep or with the pain and agony of unimaginable injuries, violence, or disease?

> We characterize Death as some mysterious black figure, hooded and carrying a scythe ready to cut down and harvest a quota of people and things as one would gather a crop of wheat. Sometimes Death moves softly and quietly, like through a baby's nursery, until he touches the one he wants. Sometimes we picture Death as roaring across a dark sky on a black charger, or rumbling across the horizon like deep thunder or a tornado, or creeping in his measured pace.
>
> Death may strike suddenly, without warning, violently: a stroke, a heart attack, an accident, a murder. When Death calls, there is no bargaining. We cannot challenge Death to a chess match with the victor claiming immortality. The most we can hope for but seldom are granted, if we are not ready for death, is a short reprieve. Many dying people, especially those dying of a disease such as cancer, have rallied during their last few hours offering false hope only to relapse and quickly sink to their death.[4]

FIGURE 8.2 Men of the 16th Infantry Regiment, US 1st Infantry Division wading ashore on Omaha Beach on the morning of June 6, 1944
Source: www.wikiwand.com/en/Normandy_landings. Photo credit: Chief photographer's mate (CPHoM) Robert F. Sargent.

There are those who pray for death but are not heard and there are those who pray to live but are selected. Gandalf in J.R.R. Tolkien's *Lord of the Rings, The Fellowship of the Ring*, summed it up as, "Many that live deserve death. And some that die deserve life." A peaceful death at an old age is what we all hope for, but in the violence-laden world of crime, warfare, racism, domestic violence, victimization, and religious confrontations, death can and will come in surprising moments of violence.

Death brings silence and an end to the pain and suffering that fear anticipates. So it is not death we are afraid of, it is the thought of bruises, pain, injury, disfigurement, and suffering. We may be concerned about the emotional pain others will experience upon our death, but our hurt, anxiety, and pain are gone.

> Tell me what your worst fears are, I bet they look a lot like mine. Tell me what you think about when you can't fall asleep at night. Tell me that you're struggling. Tell me that you're scared. No, tell me that you're terrified of life. Tell me that it's difficult to not think of death sometimes. Tell me how you lost. Tell me how he left. Tell me how she left. Tell me how you lost everything that you had. Tell me that it ain't ever coming back. Tell me about God. Tell me about love. Tell me that it's all of the above. Say you think of everything in fear, I bet you're not the only one who does.
>
> *La Dispute*[5]

The Characteristics of Risk

Risk Perception

> We sometimes fail to take adequate precautions against relatively larger risks that do not cause elevated concern [...] Risk perception helps to explain why our fears often do not match the facts. In more detail, risk perception research has found that there are several consistent characteristics of risk that form the basis of our perceptions.
>
> **1. Trust**
>
> The less we trust the people who are supposed to protect us, or the people, government or corporate institutions exposing us to risk in the first place, or the people communicating to us about the risk, the more afraid we will be. The more we trust, the less fear we feel.
>
> **2. Dread**
>
> A risk that kills you in a dreadful way evokes more fear than one that kills more benignly ... Cancer, a terrible way to die, evokes more fear than heart disease despite the fact that heart disease kills roughly 25% more Americans each year.
>
> **3. Control**
>
> If you feel you have some control over the process determining a risk that you will face, the risk will probably not seem as threatening as if it was determined by a process over which you felt you had no control.
>
> **4. Choice**
>
> A risk we choose seems less dangerous than a risk that is imposed on us.
>
> **5. Uncertainty**
>
> The more uncertain we are, the more we protect ourselves with precaution and fear.
>
> **6. Novelty**
>
> New risks tend to be more frightening until we have lived with them for a while and our experience has helped to put the risks into perspective.
>
> **7. Awareness**
>
> The more we are aware of a risk, the more we are likely to be concerned about it. Concern about child abductions rises when the press is full of coverage of an ongoing case, although the probabilities are the same before that case showed up in the papers and after it is resolved.
>
> **8. Can it happen to me?**
>
> Any risk seems larger if you think you or someone you care about could be a victim.
>
> **9. The risk–benefit trade-off**
>
> If we perceive a benefit from a behavior or choice, the risk associated with it seems smaller. If there is no perceived benefit, the risk seems larger.
>
> **10. Catastrophic or chronic?**
>
> Hazards that kill a group of people at one time in one place (such as plane crashes) evoke more fear than hazards that may take more lives, but over space and time (such as heart disease).[6]

In his dystopian novel, *Nineteen Eighty-Four,* George Orwell introduced fear to his protagonist Winston Smith, through the use of rats and "Room 101" as the worst thing in the world.

> "You asked me once," said O'Brien, "what was in Room 101. I told you that you knew the answer already. Everyone knows it. The thing that is in Room 101 is the worst thing in the world." The door opened again. A guard came in, carrying something made of wire, a box or basket of some kind. He set it down on the further table. Because of the position in which O'Brien was standing, Winston could not see what the thing was. "The worst thing in the world," said O'Brien, "varies from individual to individual. It

> may be burial alive, or death by fire, or by drowning, or by implement, or fifty other deaths. There are cases where it is some quite trivial thing, not even fatal." … "In your case," said O'Brien, "the worst thing in the world happens to be rats." … "Courage and cowardice are not involved. If you are falling from a height it is not cowardly to clutch at a rope. If you have come up from deep water it is not cowardly to fill your lungs with air. It is merely an instinct which cannot be destroyed. It is the same with the rats. For you, they are unendurable. They are a form of pressure that you cannot withstand, even if you wished to. You will do what is required of you."[7]

Seeing the rats, Winston Smith lost all emotional control, shuddering, crying with terror. At that moment he was no longer an individual, but a lover of "Big Brother." O'Brien proclaimed him "cured."

Being a Crime Victim—A Child's Experience with Fear

Fear born of trauma in childhood will often lead to fear and anxiety as an adult. Fear of violent treatment remains as a part of someone's personality even into adulthood. An emergency room nurse (now retired) tells of her experience with fear when she was a child.

> I can speak about the fear a child has when abandoned for two weeks at the age of five with three siblings younger and one a year older. Not as long as some children have been abandoned, but long enough that when what little food there was was gone, the electric power turned off, and a small fire started in a chair. I can speak to you about being sexually molested by a hateful stepfather, being dragged by the hair across a gymnasium floor, held over a bridge by my hair, the feeling of large hands around my neck, feeling the breath leave, of what it feels like when the barrel of a loaded gun touches your forehead and you know you are about to die. I can tell you the fear a woman feels when she is raped and sodomized and that death is a reality, being tied with a phone cord and held prisoner for a day which can seem like an eternity to the one who is tied. I can speak to you about feeling so afraid, so unworthy, degraded and being told over and over again, "You just cannot do anything right!"; that you can look at yourself in the mirror and say to yourself, "What's the use of going on?"[8]

That was a true story shared by the woman who lived it when she was a child. She is now a mature adult but the memories haunt her to this day. The anxieties and nightmares she has suffered since cause her to not trust men and her marriage resulted in divorce. Today she is very cautious in all her activities and remains wary, even of friends.

Bullied Child; Anxiety-ridden Adult

Fear brought about by childhood bullying experiences often lingers into adulthood, shaping one's adult viewpoint of life. This is one man's story of how fear of life has given him a cynical approach to living as he ponders peace from fear, making him relish death. During a lengthy interview with him, it was learned that he is bi-polar (manic-depressive with PTSD). The elements that carry him are anger and daily usage of medical marijuana. This man's reflections are very common in the persons who have engaged in school and church shootings. It is a feeling of revenge for being alienated or isolated and bullied.

When I was young, bullies instilled fear in me. I was too passive, too weak, and obviously naive about the evils of mankind. Progressing into elementary school, I was threatened and physically abused, though still too passive and naive to take action. In fifth grade I was made to wear glasses, which further separated me from my "peers." I already felt no inclination toward interacting with anyone. Being a "loner" further inclined them toward bullying, and the fact that I never spoke out didn't seem to help.

Over the years this pattern continued, with rage and bitterness simmering. The pattern continued throughout high school and onward, until something internal snapped. Perhaps it was my general viewpoint that life is pointless, with the bad far outweighing the good, and my overall disdain toward people by way of my experiences that constructed within me a strong apathy and negativity toward others. The experiences of losing, consistently losing, further reinforced the apathy, which welled into rage. By the time I'd grown into a full-sized man, and had enough worldly experience, the rage and apathy had grown into a monster who I came to love. Decades of sustained pondering over pointlessness and suicide eventually overcame the fear, to the point that it subsided nearly completely. Soon thereafter, I had a few altercations, with the highlight being a physical fight where I was able to destroy a bully from my past, unleashing all of my hatreds and fury onto one man with fists, glass and forearms. That experience I enjoyed every bit as much as sex.

For well over the past decade, the rational fears remain—such as harm coming to loved ones—but has entirely subsided when it comes to myself. Placing little value on my own life has meant that I've not backed down from another person or situation for a good many years, and that I haven't yet received negative results from my "fuck you" stance toward people brings further self justification. Logical fears of being homeless and destitute will always remain, but the fear of others and the fear of death is gone.

I had an opportunity to die last year, bleeding out in the street. My wife unfortunately called me an ambulance, an action toward which I still hold resentment. I look at the Earth and deduce that it isn't the planet we think that it is, but rather is the physical manifestation of the HELL people speak of so commonly. Having known that all along I feel ashamed to have fathered wonderful children who will be tested, disappointed and horrified as I have. My fear is for them, but never for myself. Many times I await and welcome my own demise. Perhaps there are patterns which have led to "PTSD." More damaging, however, has been a sustained pattern of consistent disappointment—at the world around me, the lack of insight and understanding among people, and at having made incorrect choices consistently which have irreversibly sent me down a path of sustained disappointment. The analysis of all these events combined has left me in a state of waiting to leave this planet, removed of the fear of death.[9]

Summary

Perception helps to explain why our fears often do not match the facts. See that shadow on the darkened bedroom window? Shaped like a large and moving animal? It seems to be there then it disappears. What is that black figure? Is it scratching the window? Are you filled with dread, anxiety, fear? Is it a lion escaped from the city zoo? Or is it a person sneaking across the second-story balcony seeking an open or unlocked window? Are your windows locked? Perhaps it is a psychopathic killer you read about in the morning paper? Who is the next victim? Is it coming for you? Are you awake or sleeping? Is it your imagination? Is it real? Or is it just a tree branch shadow dancing in the wind? Do you get out of bed to investigate or does fear keep you in place? Can you go back to sleep? Is it safe? What is the Unknown? Are you Afraid? Are you afraid of being raped or injured? Are you afraid of Death?

Review Questions

1. Expand upon the meaning of "Kill one, frighten a thousand."
2. Why is fear of the unknown the oldest and strongest human fear?
3. How can fear of abandonment by parents, the violence upon a small child, kidnapping, rape, and bullying have life-time consequences?

Notes

1 H.P. Lovecraft www.goodreads.com/work/quotes/124218-supernatural-horror-in-literature; Retrieved January 1, 2019.
2 www.washingtonpost.com/wp-srv/special/local/sniper-10-year-anniversary/. Retrieved July 8, 2015.
3 Santa Cruz, Nicole; "100 days, 100 nights: LAPD Deals with Fears and Rumors." *Los Angeles Times*: August 9, 2015; https://homicide.latimes.com/post/100-days-100-nights-lapd-qa/. Retrieved October 17, 2019. Quoting an LA Police captain explain how the fears and rumors were dealt with.
4 June, Dale L., *Waiting for God*; an unpublished book.
5 La Dispute is an American post-hardcore band from Grand Rapids, Michigan, formed in 2004.
6 Ropeik, David; The Consequences of Fear; http://dropeik.com/dropeik/pdf/embo.pdf. Retrieved April 1, 2017.
7 www.orwelltoday.com/readerrat.shtml. Retrieved July 15, 2015.
8 Taken from a conversation with a personal friend – Jill (last name withheld).
9 Taken from an interview with a personal friend who wishes to remain anonymous.

9

MODERN SOCIETY AND THE POLICE

Law enforcement and the criminal justice system have evolved from the "fastest gun" and the "toughest man on the block" of the 1880s to highly trained and sophisticated men, women, and technology. Society has also changed, expecting the police to be more than on patrol walking a "beat." A law enforcement officer must be skilled in numerous areas such as sociology, psychology, human relations, and much more. In modern society a police officer is many things: a parent to kids on the street, a marriage counselor, a guide, a psychologist, a coach, an ambassador representing the city, a priest, an arbiter to settle domestic disputes, an information booth, an enforcer of the law, a philosopher, a protector of victims, and many other services to all, including being available to be called when help is needed. That is the ideal!

In 1955 the Los Angeles Police Department Training Division sponsored a contest to determine a fitting slogan for the division. The winner was Officer Joseph S. Dorobek with the words "To Protect and to Serve." At the time the television program "Dragnet" was very popular and the motto began to appear regularly on the tv series. In 1963 the Los Angeles City council formally adopted the slogan as the official motto of the entire police department. Since that time the slogan has become a cliché at several police departments across the United States and is now part of police legend.

> "It is ironic that police are both a major support and a major threat to a democratic society." On one hand, police enhance democracy by exemplifying one of its central tenets, the rule of law, while also suppressing crime. On the other hand, police are granted by government the exclusive power to use force, which can be abused to undermine democracy. In either case, they play a prominent role in the success or failure of a democratic society.[1]
>
> *MIT Professor Emeritus Gary Marx*

When police officers, or officer recruits are asked "Why did you choose to be a police officer?" the number one answer is "To help others who need help but are unable to help themselves." "To Protect and to Serve" and to "help others" are noble ideals but, unfortunately, are not always adhered to by all officers.

Law enforcement today is more complicated and difficult than in the past. It's not the dangerous risks involved because that has not changed. What has changed are the attitudes of the public and the police. The public has various perspectives and viewpoints of policing. Those viewpoints, often misleading, are seen in short clips on social media or broadcast on major news outlets. The images of instances of police over-reaction can be misleading but lend themselves to a populace determined to bring allegations and claims against officers for misconduct. On the other hand, an attitude of "us versus them" mentality of the police sub-culture forms a distrustful opinion of non-law enforcement personnel.

In the current social environment, in the eyes of many in the public, police are no longer the "Officer Friendly" "here to help" or "to Protect and to Serve" ideal. There is a crisis of trust between the police and the public. The crisis has become a number one current issue in law enforcement.

After a high point in crime during the 1980s and 1990s, the new twenty-first century ushered in a brighter era of policing and community relations. With enlightened training and social awareness, the schism between police and the people narrowed to more compatible relations. Unfortunately, the "truce" began to erode and failed in the second decade of the century. The crime rate rose, and the number of police shootings of "perpetrators" increased amid growing fears and insecurities caused by increasing numbers of mass shootings in schools, movie theaters, work places, and churches.

"This is a time in which law enforcement is the target for criticism for almost everything they do and officers are constantly questioned by the public and the media without the benefit of all the facts."[2] A paradoxical dichotomy has developed in the minds of society and the police. On one branch of society's dichotomy lie the belief and determination that police have a mission of protecting the public from lawlessness and fear of becoming a victim. The second branch dictates that the police cannot be trusted, following an agenda of their own, often leading to the use of excessive force, including lethal force. Contrarily, the police paradox is that they are sworn to a mission of providing the most humane methods of preserving the peace and saving lives. Yet they have the authority to employ "reasonable and necessary" force, including lethal violence, to engage or stop a threat to their life or the lives of others. The key words in that authority are "reasonable and necessary" but it must also include "immediate threat to lives."

With the advance of social media, police activities are being scrutinized more and more, often with a biased perspective showing the actions of police in a very dark light. Alternatively, the bright light of publicity has documented several instances of police misconduct and unnecessary aggressive force, having a serious consequence of provoking fear and distrust of the police by an aroused public.

> But we're not given to revolutionary action in this country. Each generation is a little more removed from the deep-seated concerns about liberty of the generation before. We just don't seem to value privacy and freedom anymore.[3]
>
> *Donald Santarelli*

We are living in the age of everyone having a camera in their hands at any given moment, to include agencies that have adopted the use of body cameras. Body cameras are a great tool to use, in order to prove the individual officer is doing his/her job per protocol. However, with the use of body cams, a lot of discretion the officer might have had can go out the window, for fear of repercussions from the officer's agency. Police discretion is a great tool, however there is a time and place for it, and we have to trust police officers are using sound judgment when exercising their discretion.

A normal law-abiding citizen wouldn't or shouldn't have fear of the police. The individual's personality is a major factor in fearing police if he or she has done nothing wrong. If a person is a big "worrier" or just frightened by the sight of police for no apparent reason it may be because these individuals also think of the worst case scenarios and think a police officer is going to come harass them for no reason at all. Or since they were kids these law abiding citizens are taught by their parents that "cops are bad, they aren't here to protect but to do harm to us." Growing up with that mentality could definitely grow an unhealthy fear or hate for police officers.

Traditionally, children are told "Police are our friends and they are here to help us." We tell them whenever they are afraid or feel endangered they should call the police because the police will come to help them. "Wave and smile at the nice policeman." Yet, when an adult has an encounter with the police, he or she will swear at and condemn the officer. Conversely the parents who instruct their children about the ideal belief of "police are good" are also the same parents who unfortunately tell their children, "If you misbehave, I'll call the police and have them come and take you away to jail and you'll never see mommy or daddy ever again." Of course statements like that ignite separation fears in the child and an attitude of fearfulness of the police.

Such utterances can develop a fear in the child that the police will come, and they will never see their home again. It causes separation anxiety in the child, telling him the police, used as punishment, are bad, and should be avoided. Children also hear their parents praising police. However, if and when that same parent is cited for something as simple as a traffic violation in the presence of a child sitting in the back seat, he will condemn and curse the officer, call him profane names and say the officer is "no good and even stupid" after the officer leaves. This attitude is also imprinted in the mind of a child and is carried over into the teen and adult years.

Those times a child sees an officer using any force, whether reasonable and necessary or unreasonable and unnecessary, or views it in the movies and television programs and there are no consequences paid by the officer for inflicting a beating, tasering, or even killing an individual, a picture is painted in the mind of the child that police can get away with anything, including murder. Again, this leads to more fear of the police.

Police visit schools, sponsor after school programs, and conduct "explorer" training for those teenage students who may want to become police officers in the future. All of these activities are geared toward winning the confidence of young people who have been told "police are here to help us and their mission is to protect and serve."

Random comments found on social media (Facebook) on November 3, 2018:

- I find it incredibly sad how quickly so many "Americans" who once called the police heroes now call them villains. Proud to say still heroes to me.
- I find it sad that they have forgotten they are supposed to be heroes that protect and serve not cold blooded killers of innocent people. It's not the perception that has changed it's the police culture. The people I feel least safe around in the US are the police and I'm white I can't imagine how people of color must feel.
- Considering most police officers are heroes doesn't mean we shouldn't recognize that some are not.
- [Y]ou are correct there are some officers that have crossed the line but there are tens of thousands of civilians that do it every day but nobody wants to criticize the people that break the law and disrespect law-enforcement or attack law enforcement or straight up disrespect law-enforcement every day. It's always the police officers fault. When are people like you going to appreciate and highlight the millions of great things that law-enforcement

do every day to help and save peoples lives. All you do is criticize but yet it's always one-sided because you have never been a police officer or put yourself in a dangerous situation just doing your job. If you think any police officer wants to pull their weapon and take someone's life then you are sadly sadly mistaken.

- So it has nothing to do with civilians attitudes or approach towards police officers. It's clearly the police officers that wake up every day wanting to hurt people and deal with all the BS in this world trying to protect and serve. Are you out of your fucking mind. When are people like you that have nothing to do with law-enforcement going to wake up and realize that it's attitude that determines the attitude of the police officer not their skin color. It's ridiculous that people like you criticize and critique things that you know absolutely nothing about and have never been involved with. You start running in the direction of danger or getting involved in a dangerous situation to protect someone else and then you're comments will have some weight. You live in a foreign country at your own choice but yet you criticize the people that protect the greatest country in the world.
- This is the exact mindset[4] that is wrong with liberal America now ... you want to call cops killers and your scared of them but you'll call 911 on my taxpayer expense every time you need help ... I'm an insider on this issue and you'r wrong ... a few bad apples don't qualify for the good of the whole ... and do me a favor, stop making it a race issue. This line is so old, get a new method of complaint because this one is a tired one.
- If you want to know what the police deal with day in and day out, watch A&E's Live PD sometime. Maybe you'll have a better perspective on what police deal with daily. Far more cops are gunned down in cold blood, than "innocent civilians." Yes, police culture has changed. Thanks, Obama. Thanks, Mainstream Media. For fueling and pushing the anti-police attitudes and agenda that further divide our country along racial lines.
- They only love us when they need us.
- Our police, EMTs and fire fighters pick up the broken bodies after crimes, accidents and disasters. They run toward danger, offer comfort and save lives. If you find yourself in a life-threatening, terrifying situation, who do you call, BLM headquarters, or the police? Community activists do some good work, and there are problems to be solved. But send me a cop, any damn cop, if I'm in trouble.

One former officer wrote the following as part of a Christmas message to his "family" (also found on Facebook, November 3, 2018):

> As a Police Officer, I worked the streets many Christmas Eves and Days so other Officers with families especially those with children could be home with theirs. We responded to calls and kept the community safe but most importantly we handed out gifts and food to families who were less fortunate than us and changed their Christmas for the better. For my brothers and sisters in blue that were there with me, you know who you are and you will always be my family.

Somewhere in this socially constructed rupture of police/citizen relations, a subculture of "them versus us" attitude amid distrust, perhaps fear, developed. The police subculture enhances suspicion of anyone and everything not police; it's an attitude of "If you are not one of us, you are one of them." An idealistic officer, a recent graduate from a police academy, is partnered with an experienced officer who has witnessed the worst of human frailties, debauchery, and deviance in a short time learns distrust, suspicion, and an abundance of "them versus us." In turn, idealism becomes cynicism and misuse of power and influence.

On one hand, the new officer eager and ready to prove himself, who may be overly nervous to be on his first call alone, does not always make the best decisions. This can lead to a misinterpreted hand movement, and a shooting, while justified, could be avoided. Justified because the officer was probably in real fear for his life but avoidable because experience or a calmer demeanor would have led to a different outcome. On the other hand, you have the experienced officer who for all you know has been in ten shootings and could potentially have a flashback or knee-jerk reaction to a certain action, which could also lead to a shooting. Both of these scenarios depend heavily on the individual police officer and other variables such as the suspect's attitude or the previous call the officer was on. Everything in hindsight is 20/20 and nobody can ever speak for the officer in the moment that he or she pulls the trigger, and one second of hesitation could just as easily let the suspect pull the trigger.

Many sheriff's departments require fledgling officers, newly graduated from the academy, to begin their official duties working as "correction" officers in the county jail. The intent of this practice is to acquaint the rookie officer to the felons and inhabitants of the "mean streets" of the city so when the officer finally begins patrol duty he or she will have learned to distrust and be suspicious of everyone.

Fear of Police in Society

Policophobia

> Policophobia "is more commonly used as slang which means a fear of police encounters that appears to be more associated with a class of residents such as renters, low income households, young, non-white, and those with a sense of disenfranchisement from government, society and neighbour."[5]

In his book *The Most Dangerous Gang in America: The Police*, activist Richard Jeanty documents example after example of cases where Latino and African-American men and women have lost their lives senselessly after coming into contact with the police.

> As a black man, I want people to understand what it feels like to be in my shoes. I know that I am a potential victim. And it makes no difference if you are from the "hood" or live in the suburbs, in your car or in your home … Make no mistake, the message is loud and clear: no matter who you are in this society, as long as you are black, you can still be stopped, searched and arrested by the police. And all black men know, a seemingly minor interaction with the police can turn deadly … Clearly, police involved killing of brown and black people will continue until concrete steps are taken to bridge the gap between police officers and the people they are paid to serve.[6]

After a questionable officer-involved shooting there are calls for further police training; some citizens even suggest the officers aim to wound or shoot the gun out of the suspect's hand. That may work on television and in the movies, but in real life it is nearly impossible and certainly impractical and dangerous.

In a Podcast titled *5 Reasons Law Enforcement in America Creates Fear and Resentment Instead of "Safety"* by Larken Rose, originally published in *The Free Thought Project*, [7] Rose provides five reasons why he has no trust of police officers. Though this is one man's opinion addressed to *Dear Cop*, it may reflect the thoughts of many.

- **Just Following Orders**—Enforcing whatever decrees the politicians may enact, no matter how arbitrary or ridiculous they might be, and then trying to excuse their behavior using the classic cop-out, "Hey, I don't make the law, I just enforce it." … aware that when you say that, you are essentially admitting that whatever the politicians tell you to inflict upon your fellow man, you will, without question.
- **Legalized Piracy**—Despite the silly rhetoric about "protect and serve," the primary job of most law enforcers is to look for excuses to take people's money. Most police are nothing more than glorified tax collectors, always on the lookout—often as a result of "quotas"—for whatever trivial violation or infraction might justify a citation or an arrest, usually for victimless "crimes."
- **Prone to Violence**—Teaching someone that he has "authority" over others and training him to try to gain control over every situation will naturally cause someone to become an arrogant, abusive bully. If you truly believe that wearing a badge and a uniform gives you special rights, gives you the right to detain and interrogate people, boss them around and control them, then you are automatically going to treat others as your inferiors, and you will lash out at any who don't bow to your supposed "authority." … It should be clear to anyone who understands psychology and human nature why police have much higher rates of domestic abuse. Your job trained you to be a violent control freak. If and when your violence changes from "legal" violence to "illegal" violence, why should anyone be surprised?
- **Gang Mentality**—When a criminal cop abuses, assaults, or even kills an innocent person, the vast majority of other cops can be counted on to take the side of the badge-wearing perpetrator. Police unions and departments are constantly making excuses, even for outright murder, always insisting that whatever anyone with a badge did must have been justified. "It's a dangerous, stressful job, you don't understand what it's like, and you don't know the whole story, and he feared for his life."
- **Criminals with Badges**—Police are only human beings, but human beings who are trained to control and extort others, trained to view and treat others with disdain, and trained to believe that they are so important and righteous that basic morality doesn't apply to them. Add to that the fact that police "take care of their own," and rarely investigate or prosecute their fellow officers for criminal conduct, and it is obvious why so many cops become the criminals they pretend to be protecting society from. … When one officer does have the fortitude and courage to speak or act out against police misconduct, invariably all the other badge-wearers will gang up on the "traitor," for having chosen truth and justice over loyalty to the gang.

The "Other" and "Us vs. Them"

"'The Other' is an individual who is perceived by the group as not belonging, as being different in some fundamental way. Any stranger becomes *the Other*."[8] That appears to be the mindset of many police officers, establishing an invisible barrier between them and the general public. It may also bleed over into his family life. The veteran officer begins to accept professional bonding of his fellow brothers and sisters in blue as his family and everyone else is "the other." As such he wants to be portrayed as brave and fearless, willing to demonstrate his courage to the family by unhesitatingly entering into any set of circumstances involving force and violence, sometimes exceeding "reasonable and necessary" force.

It has been said many times that only people in distress or who have a need will call the police, reinforcing the concept of associating police with trouble. When a police car is seen parked in front of a business or home it is normally assumed there is trouble or a problem exists. Police will see a call for service or help as a potential threat and danger to their safety and life; thus, based on their training and experience, they approach every call with caution, alertness, prepared to use whatever force necessary. Often they will approach a common traffic stop with their weapon holster retention strap open with a hand on their gun. This is a good police tactic for safety because stopping a vehicle with an unknown driver or occupants has been a leading cause of a majority of police shootings. Looking at the same situation from the viewpoint of a driver of a vehicle, fear is created for him because he doesn't understand the reason of the officer's cautious approach, which may seem intimidating.

Police rolling "code three" (emergency—red lights and siren) to a call experience an adrenalin high and excitement induced by the moment. At the same time, anxiety, maybe dread, both lesser emotions of fear, begin to stress them; anxious to arrive timely and safely to the scene but also suppressing thoughts of self-doubt that the conditions they are about to encounter are beyond their abilities to adequately resolve. Overcoming the conditions of adrenalin and anxiety doubts, the officers may not be able to control their feelings and may release them in moments of anger and consequent actions of overly reactive violence.

Many of those over-reactions are caught and shared on social media, which often presents a distorted perspective. However, the stereotype of police reacting to racism has a history. In a song written by American folk singer Woody Guthrie on August 16, 1946, Guthrie sums up a history of police over-reaction, based on racial hatred. The actual historical incident he describes is strikingly similar to what, unfortunately, has been reenacted in many instances in contemporary times.

The song, *The Blinding of Isaac Woodard* [9] depicts the true events about Sergeant Isaac Woodard, a black World War II veteran recently discharged from the Army after three years of fighting in the Pacific. On February 13, 1946 Woodard was on a bus in South Carolina on his way to meet his wife in Winnsboro when the bus stopped in Batesville. He asked the driver if he had time to go to the restroom. The driver said, "No," and swore at Woodard and called him a derogatory name. Woodard swore back. The driver then called the police who arrived and took Woodard into custody.

The chief of police of the two-man Batesville police department, Lynwood Shull, took Woodard to jail and beat him severely in the head and face with his "billy" (a short hardwood club) or "sap" (a short leather flexible lead-loaded weapon having semi-flexible spring steel running the length. An officer normally carries them in a special pocket on the uniform trouser legs), resulting in causing blindness in Woodard. After several extensive investigations by the FBI and the National Association for the Advancement of Colored People (NAACP), Shull was indicted and charged with "aggravated assault" and violation of the civil right statute. He was acquitted by an all white jury.

The Blinding of Isaac Woodard[10]

My name is Isaac Woodard, my tale I'll tell you;

I'm sure it'll sound so terrible you might not think it true;

I joined up with the Army, they sent me overseas;

Through the battles of New Guinea and in the Philippines.
On the13th day of February of 1946
They sent me to Atlanta and I got my discharge pin;
I caught the bus to Winsboro, going to meet my wife,
Then we were coming to New York City to visit my parents both.
About an hour out of Atlanta, the sun was going down,
We stopped the bus at a drugstore in a little country town;
I walked up to the driver and I looked him in the eye,
"I'd like to go to the washroom, if you think we got time."
The driver started cursing and then he hollered "No!"
So, then I cussed right back at him, and really got him told.
He said, "If you will hurry, I guess I'll take the time!"
It was in a few short minutes we was rolling down the line.
We rolled for thirty minutes, I watched the shacks and trees,
I thought of my wife in Winsboro waiting there for me.
In Aiken, South Carolina, the driver he jumped out;
He came back with a policeman to take me off the bus.
"Listen, Mr. Policeman," I started to explain,
"I did not cause no trouble, and I did not raise no cain."
He hit me with his billy, he cursed me up and down,
"Shut up, you black bastard"; and he walked me down in town.
As we walked along the sidewalk, my right arm he did twist;
I knew he wanted me to fight back, but I never did resist;
"Have you your Army discharge?" I told him, yes, I had;
He pasted me with his loaded stick across my head.
I grabbed his stick and we had a little run, and had a little wrestle;
When another cop run up with a gun and jumped into the battle;
"If you don't drop the sap, black boy, it's me dropping that's dropping you."
So I figured to drop that loaded sap was the best thing I could do.
They beat me about the head and face and left a bloody trail
All down along the sidewalk to the iron door of the jail;
He knocked me down upon the ground and he poked me in the eyes;
When I woke up next morning, I found my eyes was blind.
They drug me to the courtroom, and I could not see the judge;
He fined me fifty dollars for raising all the fuss;
The doctor finally got there but it took him two whole days,

He handed me some drops and salve and told me to treat myself.

It's now you've heard my story, there's one thing I can't see,

How you could treat a human like they have treated me;

I thought I fought on the islands to get rid of their kind;

But I see the fight lots plainer now I am blind.

"The Blinding of Isaac Woodard" Words and music by Woody Guthrie

Community Policing

In earlier times, all the way back to the nineteenth century, police officers in towns and cities had a regular foot patrol or "beat," wherein they came to know the inhabitants, merchants, and "the mean streets." Everyone knew their local officer and he knew them by their first names. The neighborhood may not have liked him personally, but they respected him, even when respect was earned at the end of a night stick or "Billy club." Often it was respect born from fear of "coppers."

State police, or troopers, as they are called in many states, were frequently transferred to various localities and counties. The reasoning was that if a "trooper" stayed too long in one area, he would become too familiar with the people and could become corrupted; not necessarily in a criminal way, but to the point of being too lenient with people he was familiar with or related to.

> The National Advisory Commission on Civil Disorder concluded that the spark of most urban riots occurring in the late 1960s was poor or aggressive police action, generally taken in a minority community. Riots in Los Angeles, Detroit, Philadelphia, Newark, and elsewhere portrayed a disintegrating social structure, often precipitated by police action. The police were at once the cause and the solution to social unrest. Liberals saw them as the cause of problems, conservatives as the solution. The country was divided on these issues, and the police were caught between significant ideological shifts occurring in American political and social life (National Advisory Commission on Civil Disorder 1968). [...] [C]ommunity policing ... began in the mid-1980s and continues to the present. These trends stress a contextual role for the police, one that emphasizes greater interaction with the community toward the resolution of persistent neighborhood crime and disorder problem ... community policing is thought to break down the barriers separating the police from the public.[11]

> "Law enforcement society cannot work and, quite frankly, law enforcement cannot do its job if it does not have a trusting relationship with the community."
>
> *Florida candidate for governor Andrew Gillum in an interview October 26, 2018*[12]

Fear Before and Behind the Badge

When we discuss fear behind the badge we must view it from the officer's perspective. In a hostile neighborhood, an officer can feel his or her presence is unwelcomed, unappreciated, and despised. Officers view this commonality among inner city residents nationwide. They can't feel safe having no connection to the citizens of that community. It is a common occurrence for residents to not cooperate with law enforcement, even when police are working to solve a case.

This "I don't cooperate with the police" attitude is widely prevalent in the inner cities and it may bolster beliefs that the people neither care about their community nor would they care about the safety of the officers. Inner cities experience poverty, high crime rates, and lack of education and employment opportunities. Some officers may feel that the apathetic views expressed in the poorer communities are a clear indication of the mounting problems and hostility between officers and residents occurring every day. The unsaid reasoning of the officers may be, "If these people don't care for themselves and the quality of life they are living, it's easier to count them as dehumanized scabs of the world."

There is also the concern of fear of the unknown. Police officers working in areas where the majority of people are a minority may have preconceived bias. They may never have lived around minorities or had any experience with inner-city cultures and mixed races. From a viewpoint of a minority community, it may be easily concluded and believed that an officer's approach in an affluent (white) community is much different from that in a less fortunate (black) community. Many people in society view police officers as dehumanized by their training, experience, cynicism, and practices. The community members also say officers are crooked; they are seen as threats, as enemies that hide behind their badges and uniforms, while others in society have high respect and regard for the police. The reality is that police are merely human with the same emotions and feelings as anyone else. What separates police from others is their training and their passion to help others and bring peace and security to all.

Fear Behind the Officer – The Family

Lost in the public/police officer conflict is the family of the officer. It takes a special person to be married to a police officer. The unusual hours, the lost time with the spouse and/or children, the stress of the job and juggling of family responsibility. The officer may often face danger with no fear, but waiting at home may be a partner, a little girl, a son, the loves of his or her life.

When the officer leaves home for work maybe he kisses his wife, gives the children a bear hug, tells them he loves them, and will see them again in a few short hours. To his family, waiting anxiously for his return, those "few short hours" become lengthy and worry filled. They can't wait to hear his car in the driveway and for him to walk through the door and greet them with a smile and a hug. What they don't want to see and hear is another police car stopping in front of the house and officers formally walking up the sidewalk and ringing the doorbell to deliver the news the family has been fearful of and doesn't want to hear.

There are also those days and nights when he doesn't come home at his regular time and is several hours late without the opportunity to call and tell them he will be late. Too often this is the case, but the worry does not end. They worry when he is out on patrol and not safe at home. The stress on the spouse is terrifying and costly over the years.

What about the life of the officer's children? They believe he is a double hero; the father whose strength they admire and all the good he does for the people of the city. There are those times in school when other kids taunt and say unfavorable things about police and specifically about their police officer father. When they go to their social media and see comments about police as being overly aggressive, involved in the shooting of unarmed black men, and involved in corruption, other comments include, "The only good cop is a dead cop." They become fearful that someday someone will intentionally kill their father simply because he is a police officer reporting for duty. "No one seems to understand the fear that comes with being the daughter of an officer."[13]

Officers are aware they may receive a call of "Shots fired! Officer down!" The officer feels relief that she is not the one who is reported "down!" yet at the same time she knows it is one of her own, perhaps someone she rode into work with that day. She may also feel "survivor's guilt" that it wasn't her or that she was not on the scene to aid the downed officer. Another call the officer dreads and may receive is to "notify the family." The officer knows when she stops her car and walks to the front door, she won't have to say much, because the family can sense the bad news. When the news of a death must be delivered there are certain protocols that are suggested by "Dying Matters."

Breaking Bad News Guidelines

1. It's important to remember that the manner in which bad news is delivered will stay with the person. Therefore, it is best to break the news face-to-face. Even when this is not possible—relatives may, for example, be abroad—be sensitive to the impact that the news may have on the person at the other end of the telephone.
2. It may help to prepare yourself by rehearsing what you are going to say, especially when speaking to someone who may have disability or learning issues, or may not have English as their first language. The health of elderly people also needs to be taken into account.
3. Give yourself plenty of time when you are with the person, and make sure you break the news, as far as possible, in a safe and confidential setting.
4. If possible, make sure there are no interruptions. Switch off mobile phones and telephones, and turn off radios and televisions.
5. Stick with the task in hand. Use plain, simple language, and don't waffle or bring in unrelated issues as it can cause confusion.
6. In the majority of cases, people who hear bad news will only be able to take in a small amount of what is being said. So validate what they understand has happened, and encourage them to express their feelings. Gently correct them if necessary, and be prepared to repeat yourself if necessary.
7. Don't swamp the person. They may need physical space to take in this news. So leave it up to them if they want to be touched or held.
8. Don't promise anything that you can't deliver or commit to anything you can't carry through. This will destroy trust.
9. If the person becomes very distressed, and you are unable to stay with them, you may need to ask about someone you can contact on their behalf, such as a neighbor or friend, or family who live close by and can stay with them.[14]

Being the messenger of such tragic news adds to the burden of stress an officer encounters and keeps inside. You may find that delivering bad news plays on your mind afterwards. If it starts to cause you distress, do find someone to talk to about it: find professional help.

Officer Stress

Police officer stress is marked by many things, not all of which are related to potential death; dealing with the virus of crime, accidents, drunks and druggies, pimps and prostitutes, hostile environments, and a sick society. Stress accumulates from long work hours, missed meals, lack of proper sleep, working holidays and birthdays, anxiety caused by supervisors, departmental policies, and anticipation of what may await at his or her next call.

It is all a matter of perception. It is a matter of each individual officer determining what is stressful for him or her.

> What is extremely stressful to one individual may be only moderately stressful to another. The difference is nothing more than a matter of perception. Perception is commonly defined as "consciousness; a result of perceiving a mental image, awareness of the elements of the environment through physical sensation; physical sensation interpreted in the light of experience; quick, acute, and intuitive cognition; and a capacity for comprehension."[15]

Obviously no two officers deal with stress in the same ways. Some engage in heavy exercise such as running, heavy weight lifting, body building, martial arts training, etc. Others may meditate while others turn to alcohol or drugs; some engage in dangerous sports. Some officers internalize their feelings of stress and frustration only to release it all when confronting a non-compliant arrestee with extraneous aggression, force, and violent, perhaps lethal, actions.

Some social scientists have written that though police work is stressful, it is no more stressful than any other occupation. (From a personal standpoint, I believe my life as a university professor is much more stressful than it was when I was fully engaged as a law enforcement officer.)

Summary

Distrust and fear of a police officer by a child often reflects the attitude and words of the parent. Police attempt to neutralize those feelings by meetings with children in school and sponsoring special programs and community policing. It takes a special person to be the spouse of a law enforcement officer, whose family endures daily anxiety and stress waiting for him or her to come home from each duty day, afraid they will receive a call or visit from other officers telling them he or she has been killed in the line of duty. Unfortunately, officers may develop an attitude that the department and co-workers are family and may develop a mindset that fellow officers are "brothers and sisters."

Review Questions

1. Because of the recent upsurge in reported police violence and the public's perceived police use of unnecessary force, are the public's fear and perception justified or are they magnified unrealistically?
2. Do you believe a law-abiding citizen has cause to fear the police, based on the many incidents of police shootings and beatings?
3. In your own terms, discuss the "Us versus Them" attitude that might be responsible for the conflict between police officers and the public.
4. A police officer spends at least 40 hours a week dealing with some of the most horrendous, even life-threatening, events on a daily basis. You have a very close friend who is the spouse of a police officer. How do you relate, or how should you respond, to your relatives, friends, and neighbors when the talk turns to the "rotten police" who seem to over-react to what seems like a minor event?
5. What is the meaning of Policophobia? Is this a learned or acquired phobia?

Notes

1 *Why Are the Police Important in a Democratic Society?*www.reference.com/government-politics/police-important-democratic-society-f076b4e00ea6871b. Retrieved January 2, 2018.
2 Brad Carruthers, president of the Fairfax (VA) Fraternal Order of Police Lodge 77 on Facebook October 15, 2015.

3 Interview of Donald Santarelli by Radley Blako in his book, *Rise of the Warrior Cop: The Militarization of America's Police Forces*; Public Affairs Books, A member the Perseus Book Group; 2014.
4 According to Dictionary.com, the definition of mindset is this: "*A fixed mental attitude or disposition that predetermines a person's responses to and interpretations of situations.*"
5 http://health.answers.com/Q/what_is_the_word_for_fear_of_police; Retrieved October 21, 2018.
6 www.blackradionetwork.com/why_africanamerican_men_fear_the_police. Retrieved April 11, 2017.
7 Rose, Larken; *5 Reasons Law Enforcement in America Creates Fear and Resentment Instead of "Safety"*; The Free Thought Project; July 5, 2015; http://thefreethoughtproject.com/5-reasons-law-enforcement-america-creates-fear-resentment-safety/#qY2OJzCfWvtUklc7.99. Retrieved September 21, 2019. Used with permission.
8 academic.brooklyn.cuny.edu/english/melani/cs6/other.html. Retrieved 03/09/17. Italics by the author for emphasis.
9 Guthrie, Woodie; *The Blinding of Isaac Woodard*. More information about Mr. Woodard and his effect on President Truman and the Civil Rights Movement is readily available online. Of further interest is a 1946 commentary by Orson Wells: www.youtube.com/watch?v=P11sW1sXNbs.
10 Guthrie, Woodie; *The Blinding of Isaac Woodard*. More information about Mr. Woodard and his effect on President Truman and the Civil Rights Movement is readily available online.
11 http://what-when-how.com/police-science/community-oriented-policing-history/. Retrieved September 23, 2018.
12 "Pod Save America." *The Washington Free Beacon*; www.foxnews.com/politics/andrew-gillum-police-going-too-far-if-they-pull-out-gun-baton-or-stun-gun?fbclid=IwAR3NMGsoT0BKQDezPr7f8R4xLqc1s_DzZF698pf2rSG_i60i_MshtD5icZ0. Retrieved June 27, 2019.
13 Ray, Bri; *Protected, But Scared: What It's Like To Be A Police Officer's Child*; www.elitedaily.com/news/world/protected-scared-like-police-officers-child/960548. Retrieved June 16, 2018.
14 Telling others about a death; www.dyingmatters.org/page/telling-others-about-death. Retrieved October 7, 2018.
15 Copes, Heith; *Policing and Stress*; Pearson/Prentice Hall; Upper Saddle River, NJ; 2005.

10

VICTIMS AND FEAR OF VIOLENT CRIME

> Violence can occur anywhere, any time, reminding us of our vulnerability to dedicated predators seeking targets for personal revenge, pleasure, profit, or twisted political motives .... Although studies have found that women and the elderly report higher levels of fear of crime than do men and younger people, these two groups are much less likely to be victimized by crime. In other words, those who are most fearful actually report the fewest victimizations. The expectation as to who is fearful and who should be fearful of victimization is referred to as the fear–victimization paradox.[1]

Nothing can be true and false at the same time in the same way but a paradox may come close. A paradox is defined as "a statement that appears to be self-contradictory or silly, but which may include a latent truth."[2] Some people will always be overly cautious about everything in life. Some are more fearful of potential violence than others for various reasons. They are ill-prepared physically and psychologically to deal with any violence (as previously discussed). They have the mentality that violence is around every corner, putting them in the "sheep mentality."[3] The sheep mentality means not thinking for oneself but simply following what others think or do. They are highly influenced by outside sources such as the news media, rumor, gossip, and the fears of others. They rely on the police or other public or private security services to give them the feeling of security they need; while at the same time they grasp on to the idea that the police will not respond in a timely manner or will fail to provide the peace of mind their worrisome emotions demand.

People most fearful of crime do not reside in crime-ridden areas, they are living in locked-down apartments, gate-guarded communities, installing alarm systems, surveillance cameras, buying guns; more fearful because of what they do not know than what they do. Most people that live in areas where they are more likely to become victims are aware of that fact but not fearful of it.

Life experience can reduce a person's fear of becoming a victim of crime. An example of this is career criminals or gang members who are so familiar with the criminal environment, they have almost no fear of the consequences of their lifestyle. Some of the factors that would lead certain people to be more fearful of crime than others could include past trauma of being victimized, being witness to crime, or having second-hand knowledge of a victim of a crime, i.e. news or other mass media.

The fear of crime in any society is as damaging as the act of crime itself. The typical questions run through everyone's heads, but only if you let them. Those questions are as follows: who, what, when, where, why, and how? These are essential questions to any situation. First is "who": each direction can go more than one way, such as who will be the target or who will be the person committing the act of violence? This leads us to "what": What time? What place? What neighborhood? Third we have "when." If not today, then maybe tomorrow; if not tomorrow, then maybe next week. Fourth, we look at "where": another school, a town square, backyard, grocery store, or ghetto. The most important question is "why?" Why did you choose to act in that way and why did you choose to victimize me? Lastly, we all are left questioning "how?" How did it end up this way?

> Fear is an absence of strength in a particular situation.
>
> *Robin Sharma*

It can be emotionally stressful for people who live in fear. The fear of crime can negatively affect the residents' behavior, reduce community organization, and deter new businesses from wanting to open in the area for fear of being robbed. This adds to the economic woes of an area heavy with crime. The law enforcement organizations of high-crime areas generally utilize an increased visibility, but this often backfires in low-income areas, causing the population to see the police as an occupying enemy.

Other problems include the fact that victims of crime must deal with increased fear and trauma afterward. Even for those who have not been a victim of a crime, fear can vary depending upon the demographics. Age is a factor because the elderly are more vulnerable; teenagers might fear that committing crime is the only option they have in order to fit in or get by. Gender is a factor: women express higher levels of fear of crime because they are always seen as the weaker, more vulnerable, sex and of course the risk of rape has a high fear factor. People living in non-white neighborhoods are more fearful of crime because they might not have the same income levels to upkeep the neighborhood. Crime can even change the appearance of neighborhoods such as fences, alarm systems, and bars placed on windows and doors to keep the violence out.

When it comes to violent crimes in society, everyone has some level of fear because there is always the threat of having a violent encounter. The fear of violent crimes encourages people to ensure their doors and windows are locked before they go to bed and many people have some sort of weapon nearby for protection in case there happens to be an intruder in the middle of the night. Most people make sure to lock the doors in their vehicle, lessening the possibility of being car jacked.

The amount of violent crimes shows no sign of slowing down in any areas of the United States. No matter if it's the city, a peaceful suburb, or a ghetto neighborhood, violent crimes always find a way to strike fear into people. Having said that, people do seem to be much more aware of their surroundings than they previously were. There are more neighborhood watches, more diligent people watching what's happening around them and reporting it on a regular basis. There are even apps developed to help people know what's going on in their neighborhood and the surrounding areas. This is a move in the right direction and makes criminals less likely to hang in the neighborhoods that seem to have more people constantly watching.

The other fear people have is about the types of violent crimes occurring. Violent criminals seem to be less deterred by jail time and will commit more heinous crimes without giving it a second thought. There is no fear of being arrested in some criminals which in turn makes them a little more dangerous. That, to the law-abiding citizen, is terrifying. When society has the perception that criminals are more daring, a sense of being anxious all the time is created. It can also create a hyper-sensitive reaction to surroundings, which makes people slightly paranoid at all times. It's the sad reality of the state of the world we live in, but it's a fact.

Violent crimes are so feared because they are in every part of our lives, no matter how hard we try to protect ourselves. There are kids committing violent crimes against other kids and even their parents. Some kids try to reenact the villains they watch in the latest movies, or go on a shooting rampage because they grew up doing it on video games. Violence is a learned and taught behavior, seeing it over and over on the news, television, video games, and movies. It's also an environmental issue. If a person is raised in a violent family or a family that struggles with drugs or violence, he or she is more likely to follow suit and struggle with the same issues. Unfortunately, it's like a drug: getting addicted and sucked into that world with never an easy way out.

There are reasons there is so much fear in our society. There are gangs and large groups who are notorious for crimes and violence. They are feared by everyone. They run by their own rules and, oftentimes, innocent bystanders are in the wrong place at the wrong time and end up in a victim situation.

Who is a Victim of Crime?

To a casual observer the victim of a crime is the person who has been injured, killed, or suffered a loss. In a violent crime such as rape, assault with a deadly weapon, or murder, there are several victims as a result of the one act or the decision by the perpetrator. The effects of the crime are similar to throwing a stone into a pond of water. The ripples cause consequences far beyond the initial act. The consequences of one person's decision have many ramifications, affecting many others beyond the initial victim. Coincidently when a police officer involved shooting occurs, the ripple effect is wider than the act itself.

Consequences of Being a Victim

The below diagram (Figure 10.1) illustrates the many consequences of a crime and the many victims of a singular crime. Naturally there are physical injuries, and possible death to the initial victim. If the victim is not killed there may be serious physical and mental damage, taking years, if ever, to recover from. The victim's family, circle of friends, and employer will mourn and be affected by the victim's injuries and/or death. There will be medical and/or funeral, and possibly legal, expenses to consider as well as mental trauma requiring therapy and counseling.

The individual(s) responsible for the injuries to the core victim will become a victim of his decision to victimize when he is arrested. He will encounter legal expenses, perhaps his freedom or life; his family, circle of friends, and employer will lose his companionship, and love.

First responders like police, medical personnel, and witnesses may suffer post-traumatic stress disorder (PTSD) from being involved in a horrific and sickening scene of blood, torn bodies, or other conditions of inhumanity.

Fear at Home—Domestic Violence

Fear and violence are very common in "the street." But they also reside in the safest place of all—in the home. In what was once defined as a private family matter is now better described

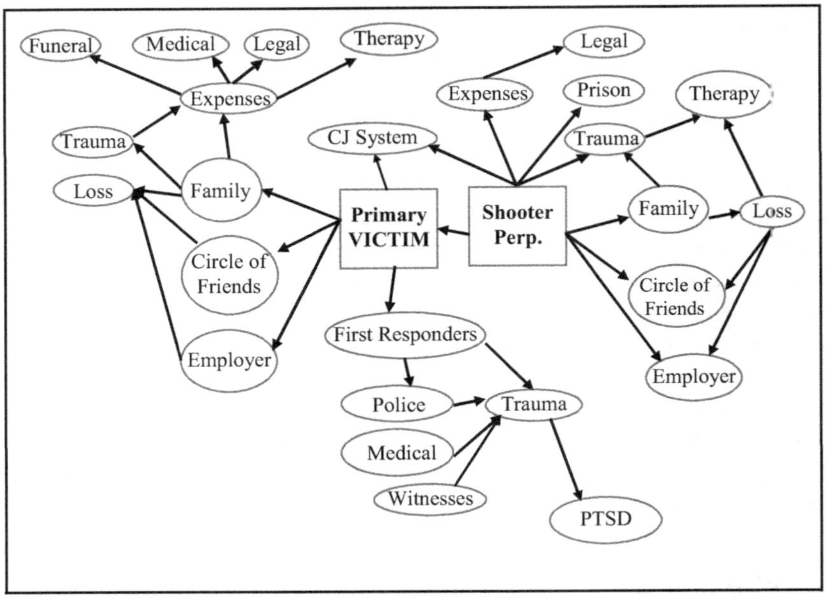

FIGURE 10.1 Who is a victim of crime?

as "domestic violence." A very common, perhaps the most common, call for police assistance is to a domestic violence incident. "Intimate partner violence" is frequently defined in terms of the abusive exercise of power and control by one partner in a relationship over the other.

Police are disadvantaged when responding to a domestic violence call. There are many unknown potential scenarios they will be facing. Who is involved? How many people are present? Is alcohol a factor? Is a gun or other weapon present? Have shots been fired? Emotions at the scene are running high and wild and often involve a weapon—a gun or knife. An officer, with his or her weapon holstered, enters a house or apartment to be met by an enraged husband, wife, or domestic partner holding a gun. Before the officer can assess the situation, the gun holder begins to shoot. In 2017, more officers were shot responding to domestic violence than any other type of firearm-related fatality, according to the National Law Enforcement Officers Memorial Fund.[4]

Unfortunately, many women are fearful of calling the police. Reasons for this include: fear of further violence from their partner; fear the police will do nothing; fear the police will arrest her if she is involved with drugs; fear she will be deported if she is an undocumented resident; fear her children will be taken away from her, and any other number of reasons.

What if the source of the violence in the home is a police officer? Abusers who are law enforcement officers have unique opportunities to assert that power and control in their intimate relationships by misusing their training, skills, experience, and knowledge as well as the power and authority inherent in their position. Officers who carry that authoritarian character and demeanor into their homes may also expect their intimate partners, children, and other family members to comply with their demands. Family members who argue, resist, or ignore them may be seen as a direct challenge to that authority. For an officer who is an abuser, those challenges may be met with verbal and emotional abuse or physical violence.[5]

The victim of an abusing police officer faces an insurmountable task in efforts to escape her abuser. He threatens her with fear of his possession of guns and skill of hitting where the bruises won't show. He also has resources at his disposal to track and locate her if she is able to find the courage and means to escape her circumstances. Even if she assumes another name and relocates to another state or country, she will remain fearful that one day he will find her and "punish" her for daring to leave him.

> One survey of 728 officers found 40 percent of the officers reported having "gotten out of control and behaved violently against their spouse or children" during the previous six months. In a second study researchers asked 385 male officers from several agencies in the Southwest whether, in the past year, they had pushed, grabbed, shoved, slapped, kicked, bitten, hit, thrown things, choked, strangled, beaten up, or threatened to use or used a gun or knife against their spouses. Twenty-eight percent of those officers admitted having engaged in some form of such conduct within the previous year.[6]

A victim of domestic violence lives in constant fear. "Fear is a tyrant and a despot, more terrible than the rack, more potent than the snake."[7] Domestic abuse is primarily about power and control; intimidation to bring fear and obedience to the victim. Police officers, firefighters, professional athletes, military personnel, politicians, and others in positions of power and control in their professional lives don't always disengage their professional persona upon arriving home.

A diagram depicting the cycle of domestic violence was developed in 1979 to illustrate the never-ending abuse and fear an abused victim in a relationship suffers. The batterer completely dominates the life of his victim; she lives "on pins and needles," never knowing when he will verbally abuse or physically attack her.

> The Cycle proposes that domestic violence starts with tension building in a relationship before an "explosion" occurs—the actual incident of abuse—followed by a "honeymoon phase" where the abuser is apologetic, even romantic, promising the abuse won't happen again. Soon after, the tension starts building again and the cycle repeats itself indefinitely. [...] "We don't label our experiences accurately. American society defines power in violent ways and then socializes men to be 'powerful,' and [says that a] "man's way" of doing things

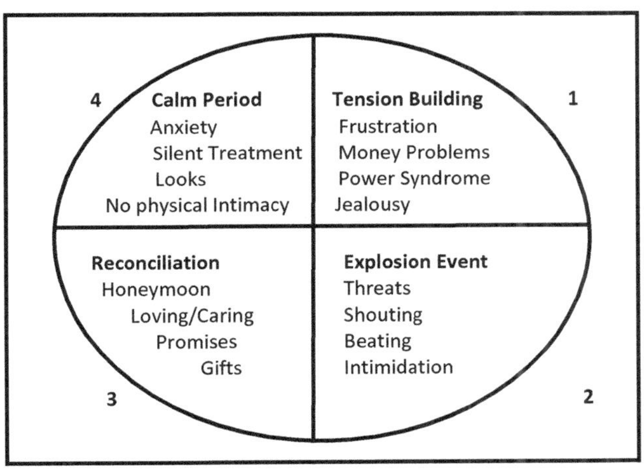

FIGURE 10.2 The cycle of domestic violence

> is superior to women and women's ways. Things that are feminine by nature are minimally undervalued in American society, if not outright seen as inferior and worthless … there are some batterers who only beat up a victim once. After that, they just have to give a look. … After someone beats you up, they can [be intimate] with you, but you're not in a position to say no. You're traumatized, shocked … so, it could be more like rape."[8]

> Does it seem that fear and domestic violence coexist? They do. One of the most powerful weapons of abuse is fear. It always starts with words—always. Maybe innocuous at first, you might be told you had a dumb idea, or how ugly and undesirable you are, followed by the comment "Nobody else is ever going to want you."
>
> Have you experienced "the glare" that comes with fear and domestic violence? You know not to open your mouth. A simple look in your direction is enough to scare you into silence. You might be grabbed by your arm, or shouting may begin. For those who are experiencing such aggressive behavior, you are probably in that box of fear and domestic violence. Ah, but you don't *think* you are a victim because, "He hasn't hit me." No, not yet.
>
> What about all the threats? Do you think they are simply idle? Most likely the aggressor will follow through by putting action behind those words. Grab your hair. Slap you in the face. Take your neck in their hand and squeeze until you feel as if all the blood in your body has pooled inside your head. That usually works as part of the fear and domestic violence card batterers like to play. If they can strike fear into their prey, they can maintain the power and control that keeps them going. You may begin to recognize narcissistic behavior being displayed. It's never their fault. They won't accept responsibility for their own actions. You *drove* them to drink, so it's *your* fault. Every flaw is cast upon someone else to keep it from appearing they are imperfect. But *you* see and know the *real* person—the one who finally trapped you into fear and domestic violence.[9]

Brenda Hill, (founder and) co-director of the South Dakota Coalition Ending Domestic & Sexual Violence, says she "would rather survivors and others looking to understand domestic violence look to the Power and Control Wheel," which was developed by the Domestic Abuse Intervention Project in 1984 after interviewing hundreds of survivors. "We documented the most common abusive behaviors or tactics that were used against these women. The tactics chosen for the wheel were those that were the most universally experienced by battered women."[10]

Women as Domestic Abusers—Men Living in Fear

Domestic violence is not limited to men abusing women. Men are also victims of domestic violence. However, they are very reluctant to report the incident to the police for various reasons. These include the fear that they won't be believed and, if children are involved in the domestic setting, the fear that their children will be taken away from them.

> "The violence really began in our family about 10 days after (my wife) realized that she had all the power [financially]. I knew I had to get my kids out. I called the largest domestic violence shelter agency in the county several times. They told me, 'Men are perpetrators of domestic violence; women are victims of domestic violence,' and hung up.
>
> I had no way out. I had no money. Whenever we bought a car, she insisted that the car be in her name only, so that if I took it and went to the movies without her approval

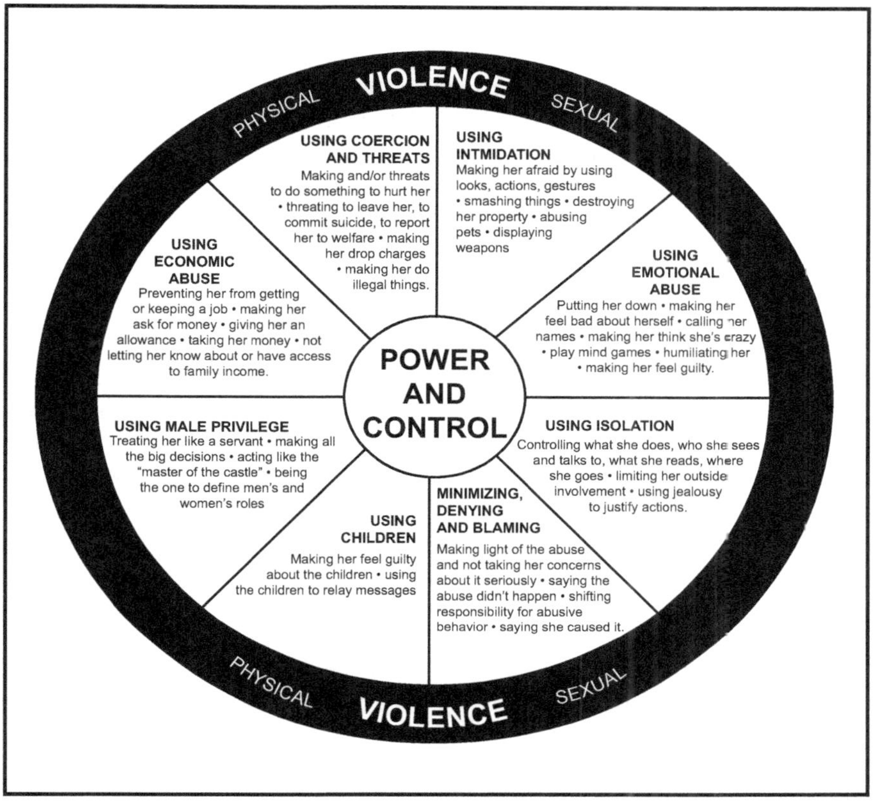

FIGURE 10.3 Power and control wheel
Source: Domestic Abuse Intervention Programs, Duluth, MN.

she would call the police, and report, 'I'm estranged from my husband, and he stole my car.' She did that several times."

One daughter says, "No one would help. Teachers, parents of friends, anyone I tried to talk to about what was going on at home told me I didn't understand, that my mother couldn't possibly be the violent party. When the police came to our home, they would always be ready to arrest my father, sometimes putting handcuffs on him. It was up to me to scream as loud as possible that it was my mom and not my dad, so they wouldn't take him away and leave me alone with her."

Domestic violence committed by women against men is generally ignored or minimized, yet more than 200 studies have found that women initiate at least as much domestic violence (DV) against their male partners as males initiate against their female partners. Research shows that men comprise about a third of DV injuries and deaths. Women often compensate for their lack of physical strength by employing weapons and the element of surprise.

The most recent large-scale study of DV was conducted by Center for Disease Control and Prevention researchers and published in the American Journal of Public Health. The study, which surveyed 11,000 men and women, found that according to both men's and women's accounts, 50 percent of the violence in their relationships was reciprocal (involving both parties). In those cases, the women were more likely to have been the first to strike. Moreover, when the violence was one-sided, both women and men said that women were the perpetrators about 70 percent of the time.

In (a) study of male victims of domestic violence, 64 percent of the men who called a DV hotline were told that they "only helped women," and over half were referred to programs for male perpetrators. Overall, only 8 percent of the men who called hotlines classified them as "very helpful," whereas 69 percent found them to be "not at all helpful." Worse, when an abused man called the police, the police were more likely to arrest him than to arrest his abusive female partner.[11]

Summary

Fear of violent crime is endemic in society. People most fearful of being a victim are the least likely to be victimized. They are the homebound elderly and disabled. Experience with danger and threatening elements shows a lesser degree of fear of violent crime than those having no exposure to actual violence. Seeing reported violence on television and social media causes a false feeling of anxiety and fear of being victimized.

Being a victim of a crime has many "shadow" victims beyond the initial victim. As a pebble thrown into a pond of water causes ripples, consequences of a crime affect family, friends, employers, first responders, etc. Consequences include legal, medical, and, perhaps funeral expenses.

Domestic violence follows a cycle and is mostly about power, influence, and domination. Men are also victims of spousal violence and battery but don't receive the same consideration as women.

Review Questions

1. Explain the fear–victimization paradox.
2. Make a list of victims of crime beyond the principal victim.
3. Why are men as domestic violence victims seldom recognized?
4. What is the difference between the cycle of violence and the power cycle?
5. Using the "Who is a victim of crime?" diagram as a model, how does an incident of workplace violence create victims beyond the immediate victim(s)?

Notes

1 Meadows, Robert J.; *Violence and Victimization*; 3rd Edn; Pearson-Prentice Hall; Upper Saddle River, NJ; 2004.
2 //https:literarydevices.net/paradox/. Retrieved January 18, 2019.
3 Grossman, Dave. "On Sheep, Wolves, and Sheepdogs." http://mwkworks.com/onsheepwolvesandsheepdogs.html. Retrieved January 2, 2019.
4 Schreyer, Natalie, "Too Terrified to Speak Up: Domestic Abuse Victims Afraid to Call Police." *USA Today*; April 9, 2018.
5 Garvey, Teresa M. JD; "The Highly Trained Batterer: Prevention, Investigation and Prosecution of Officer-Involved Domestic Violence"; *Strategies: The Prosecutor's Newsletter on Violence against Women*; Issue #14, February 2015. Retrieved March 17, 2017.
6 Ibid.
7 Wallace, Edgar; *The Clue of the Twisted Candle* (1916) www.azquotes.com/quote/522797.
8 Hill, Brenda; *Is the "Cycle of Violence" Outdated? Advocates Say Look to the Power and Control Wheel to Understand Domestic Violence*; www.domesticshelters.org/domestic-violence-articles-information/is-the-cycle-of-violence-outdated; Retrieved December 15, 2018.

9 https://carolynhennecy.com/domestic-violence-awareness/2015/fear-and-domestic-violence/. Retrieved July 8, 2019.

10 Hill, Brenda; Ibid.

11 Sacks, Glenn, M.A. and Ned Holstein, M.D; *No One Believed Me; When Men Are Victims of Domestic Violence*; www.coursehero.com/file/7019592/Domestic-Violence-Article/. Retrieved December 19, 2018.

11

FEAR AND THE NEWS MEDIA

> If it bleeds, it leads!
>
> *Television News Directors*

Fear of violence is totally justified in society. It is unfortunately part of the human condition that we act out and are sometimes violent. Fear of violence is a basic instinct and is what helps to keep a civilized society acting civilized. After all, fear is a great motivator. Because of the availability of information at our fingertips today over social media and the Internet and television, the more violent incidents will be quickly reported (and repeated over and over in every 24-hour news cycle), exposing society to more violence than could be found in an MMA free-for-all or a movie by Quentin Tarantino.

Is the public's fear of police justified? Police deal with thousands upon thousands of potential deadly scenarios across the United States on a daily basis and of course there are some "rogue officers" giving law enforcement a bad reputation but that is not how the police should be viewed as a whole. It may not be the news media's intent but, too often, they portray and spin every situation to get more viewers, which in turn makes things look racial and bad for the police. For example, if two shootings were to happen the same day in two different states— one a black officer shooting a white suspect and the other a white officer shooting a black suspect—the media would play the stories very differently.

One story would likely read "Officer shoots fleeing suspect," while the other would read "WHITE police officer shoots BLACK suspect." The two could be describing the exact same situation but because one report is labeled with race, most people today tend to see a trend of police racism without taking the time to gather the facts and develop their own opinion.

There is reason to be afraid of violence that happens in society, but it is magnified by the media. It is not necessarily the media's fault; people see the violent acts happening from the comfort of their own home and just take the media's word for it without doing their individual research of why that violent attack happened. Was it justified? Is it limited to a certain area? Is this likely to happen in my neighborhood? Yes, violence can happen anywhere and at any time but people have to stop being fearful by watching the worst of the worst violent acts that happen in this country.

Exploitation of Fear by News Media

Nowhere is fear more promoted than in the news media. There is a saying in television newsrooms, "If it bleeds, it leads." That is indeed cynical, yet real. Those types of stories and the method of delivery in a "day of doom" voice and manner are designed to bring anxiety and fear to the news watchers and to hold them captivated. It's not what is said, but how it is said. "Fear-based media" sells newspapers and advertising, garnering a large following of watchers.

It's been said that fear-based media has become a staple of popular culture. The distressing fall-out from this trend is that children and adults who are exposed to media are more likely than others to:[1]

- feel that their neighborhoods and communities are unsafe;
- believe that crime rates are rising;
- overestimate their odds of becoming a victim; and
- consider the world to be a dangerous place.

The elderly, shut-ins and physically disabled are the most vulnerable to fear from dramatized news delivery and crime-enhanced stories. They sit in their bedrooms or living rooms and watch the news all day long, seeing only that the world, especially their city, is a dangerous place. They keep their doors and windows locked with the curtains drawn, afraid to go outside. It's ironic that those who are most frightened by what they see and hear from the news media are those who are least likely to be violently victimized. Conversely, the median age of violence victims is between 18 and 25. Coincidently they are also the ones most likely to commit violence.

> The people will believe what the media tells them they believe.
>
> *George Orwell*

> The media's the most powerful entity on earth. They have the power to make the innocent guilty and to make the guilty innocent, and that's power. Because they control the minds of the masses.
>
> *Malcolm X*

Police high-speed chases attract major news media attention on live television. The chase itself is an example of adrenalin-boosting action for the chasing police, the news reporter "high above the scene in Sky Chopper 2," and the viewing audience. It's not the chase that intrigues people but the element of suspense of how the chase will end and what happens next. Pursuing officers experience high adrenalin downloads from the chase and, not surprisingly, the fear of the unknown of what to expect when the chased vehicle is forced to a stop. The fear has, in some cases, translated into violence as a release for the officers; perhaps they will react with violence toward the fleeing individual. The watching audience has an interest in the chase for how it ends, anticipating violence, but they are also fearful and looking forward to what may happen during the chase; perhaps there will be a collision with injuries. Whatever happens is carried into homes by a live news media feed with reporters analyzing and speculating on what they see.

No one in the news media has interviewed drivers and passengers in other vehicles along the high-speed chase route to determine their mental and physical response to a very reckless driver being pursued and the following police cars with flashing lights and wailing sirens. It is not difficult to imagine the fear those noninvolved drivers must feel as they see the action right before them as they also try to maneuver through heavy traffic all attempting to dodge the pursued and the pursuers. It is easy to speculate that every person interviewed would express some type of fear and would perhaps mention the bright lights of the news helicopters following the story live as the lights momentarily blinded them causing them to fear a collision with other blinded drivers.

The following section was written by Gavin de Becker of Gavin de Becker & Associates in Los Angeles. It is quoted in its entirety with Mr. de Becker's permission. The essay, written with a dark sense of humor could be considered satire and is an outstanding example of how words are used to "grab attention" even if the words ring "hollow" with insincerity.

"Media Fear Tactics" by Gavin de Becker[2]

It would be interesting if the standards of Truth in Advertising were applied to television news as they sometimes are to television commercials. In that unlikely situation, TV news writers would be required to use phrases and words that convey accurate information—as opposed to the phrases and words they use today.

I want to help you break the code of alarming newspeak so that you can more easily find the valuable information that may (or may not) be part of a story.

Given the disturbing reasons we've all been watching so much TV news, it would be understandable to overlook the sheer ridiculousness that is inherent in some of the sensationalism. Occasionally, the way TV news is delivered can be downright funny, and indeed, the ability to laugh at something indicates that we are beginning to gain perspective on it. Accordingly, some of what follows is funny, and I have a very clear purpose in offering it: I want to help change your experience of television news, help you actually watch it differently. I want to provide some tools you can use to ensure that when you watch TV news, only actual information gets through.

Though not offered as a comprehensive glossary, here are some examples of words and phrases I think you'll quickly recognize:

POSSIBLE: As in "Next Up: Possible links between Saddam Hussein and tooth decay ..."

The word "possible" doesn't really have the specificity one hopes for in journalism, given that it is completely accurate when applied to anything anyone can possibly imagine. "A possible outbreak of ..." means there has been no outbreak. "A possible connection between memory loss and the air you breathe ..." means there is no confirmed connection.

"Officials are worried about possible attacks against ..." means there have been no such attacks.

Anytime you hear the word possible, it's probably not happening right now.

LINKS:

"Next up: Possible links between convicted murderer Charles Manson and yesterday's traffic jams in the downtown area."

Are these two things linked? Absolutely, if you loosen your criteria enough, everything is linked by its presence on the same planet at the same moment in time—but only a very few links are instructive or meaningful.

Links are a great news trick, because you can tie a remote, unconfirmed, or even unimportant story to something that's really pushing buttons. "Next up: Possible links to Bin Laden" is all you have to say to get attention these days.

Almost always when you hear the word link, there is no confirmed link.

"OUR NATION'S …"

"… our Nation's water supplies …"
"… our Nation's roadways …"
"… our Nation's shipping ports …"

They use this trick to imply some large scale to a story. "A new threat to our nation's water supplies" won't be a threat to our nation's anything. Our nation is enormous. Nothing, not even nuclear bombs, poses a threat to all of any system in our society at the same time. When they say "our nation's" anything, they are usually trying to give grand significance to something that doesn't have grand significance. We might not perk up as much if they said, "A new threat to Klopp County's water supply …" The incident in which old Doc Ames' truck leaked oil into the reservoir just isn't gonna scare up enough ratings. But it could: "Next up, a new threat to our nation's water supply. An alarming incident that experts say could happen anywhere!"

SHOCKING NEW DETAILS:

"Shocking new details when we come back." Well, first of all, the details are not likely to be new, and if so critical, why are we waiting till after the commercial, and anyway, what does shocking mean at this point? Unless the news anchor reaches through the screen and pulls my hair, I don't imagine he could shock me. They've ruined another word for themselves.

AUDITORS

OBSERVERS

ANALYSTS

INSPECTORS

LOOPHOLES

"Auditors cite loopholes in security at our nation's libraries."

That's right, anytime you have an audit or an inspection, you're going to find something. Auditors are people who've been hired to write reports identifying deficiencies. Have you ever heard of a one-line audit report? "The auditors didn't a find one damn thing that could possibly be improved." Did you ever hear of an inspector who said: "We've wasted six months on this inspection, because the place is bloomin' perfect. Whoever's running this show sure thought of everything," through the loophole – but that is not necessarily so. They tell you (and the terrorists) about the loophole because it is frightening, not because it's enlightening.

"IN A CAREFULLY WORDED STATEMENT …"

"In a carefully worded statement, the President said …" Is this as distinct from those statements that world leaders just have the kids throw together? "Carefully worded" is often used to imply that something is being hidden.

SERIOUS...

"Officials consider the threat to be serious." Is that to distinguish this threat from the threats they laugh about over lunch? Taking something seriously does not mean the risk is great or imminent. It just means officials are doing what anyone would do.

"Officials here are taking no chances when it comes to school safety." Sort of. More likely, they're taking no chances that reporters will broadcast a report accusing them of taking chances.

OFFICIALS ARE CLOSELY MONITORING ...

Implies that something is imminent, and worthy of being closely monitored. "Closely monitoring" is like "Officials are on the lookout for ..." Both phrases suggest that something bad is surely coming, as if officials are standing outside looking around with binoculars.

COULD

PERHAPS

POTENTIAL

MIGHT

"NASA reports that a large piece of space junk—PERHAPS as big as a freighter—COULD enter the Earth's atmosphere sometime tonight over North America. Experts warn that it could potentially slam into the earth."

What are we to do with this report? Move a little to the left or right? They don't say, of course, that every night, thousands of pieces of space junk enter the Earth's atmosphere and completely burn up before ever hitting the ground, or that no person on Earth has ever been struck and killed by a piece of space junk. Or that if something's as big as a freighter before entry, it might end up as small as a grain of sand—but it could potentially hit your house, I suppose.

AN ALARMING PERCENTAGE ...

15%, 20%, 25% ...

"15% of Americans are at risk of being seriously injured in car accidents on our nation's highways this year." Whenever you see a percentage cited, reverse it and think about the other share in the equation. For example, from the story above you can conclude that 85% of Americans are not at risk of being seriously injured in car accidents this year. Sort of good news, all things considered. Also, phrases such "a sizeable percentage," or "an alarming percentage" can be applied to just about any percentage. Get the actual number, and then you decide if it's sizeable or alarming to you.

AS MANY AS:

"Experts warn that as many as 25,000 people in America may be carrying the deadly gene ..." or "As many as twenty states may be susceptible to radiation leakage disasters."

"As many as" means somewhere between zero and the number given.

"IN A DEVELOPING STORY ..."

A phrase used when they don't really have the story yet.

FORMER EMPLOYEES:

"But one former employee at the doomed refinery reveals shocking new information ..."

What does he reveal? That they fired him because he was too ethical, or because they didn't want to hear the truth? Or that he knew all along? Anyway, he wasn't there the night of the fire, so is he the best source of information? Truth in advertising would require the reporter to say: "We interviewed one man who hasn't been to the refinery in three months—his opinion, next."

LANGUAGE FROM ONE STORY BEING USED IN ANOTHER:

As certain words and phrases become symbolic or evocative from one type of story, they'll use them in another. In the days after 9/11 I saw a TV news report about a tropical storm making "a direct hit" on a tiny coastal community, as if the hurricane were aiming. (And the word tiny is used because it implies vulnerability. Storms that make direct hits on tiny places are frightening bullies.) A story about a flight that experienced extreme turbulence is headlined "Terror in the Sky."

DEADLY:

As in the popular "deadly virus"; this word is used to imply that everyone who gets the virus perishes, when the truth is that very few people die from the virus. If a really serious virus ends up being fatal for 20 percent of the people who contract it, then truth in advertising would require language such as: "Next up, a local man is stricken with a highly survivable virus."

It's quite a bit shy of deadly when someone tests negative for anthrax, yet in the weeks after 9/11, even a negative test for a "deadly" virus was presented as a frightening thing.

To put this into perspective, flu-related disorders killed 5000 times as many people as anthrax in 2001. Is anthrax still scary? Yes, and all the more so because of the implication that it was everywhere (colored maps showing the places in the U.S. where anthrax was found or suspected). It wasn't everywhere. Reports were everywhere. And the same report repeated seventy-five times is still the same report. But you wouldn't know that by the excited delivery: "New details emerge in that anthrax case." Details maybe, but not new—far more likely when you watch TV news, they'll be the same "new" details for the tenth time that day.

A storm is described as deadly: "We'll have new information on that deadly hurricane that's heading up the coast." A hurricane qualifies for the word "deadly" when someone, somewhere on the hurricane's round-the-hemisphere journey dies as a result of the storm. That does not mean the hurricane tries to kill all people it encounters, but that's the implication—that something dangerous is coming. You'll note that the people who die are usually in a situation far different than yours: They are on a small fishing boat at night off the coast of Peru, and you're at home 1200 feet above sea level.

"IN A LAST MINUTE DEVELOPMENT …"

"IN A SURPRISE DEVELOPMENT …"

Usually means they didn't get a news crew there in time. Or they didn't warn you about it yet, which actually is interesting, since there's only two or three possible awful outcomes involving human beings they haven't warned us about yet.

DISTURBING QUESTIONS:

As in "Disturbing questions have been raised about the safety of our nation's …" Yes, the questions are disturbing. They're disturbing everyone. Please stop raising them.

"A NEW STUDY REVEALS …"

"A NEW REPORT WARNS …"

"EXPERTS FEAR …"

"EXPERTS WORRY …"

Yes, reports and experts do seem to warn, fear, and worry a lot.

"EXPERTS SAY IT'S JUST A MATTER OF TIME BEFORE …"

They sure do.

"BUT NEW YORKERS FEEL …"

Global conclusions drawn from man-on-the-street interviews represent literally nothing. You can edit a story into "New Yorkers feel terrified," or "New Yorkers are ready to move on"—and it all depends upon which of the five interviews you cut into the piece broadcast.

Here are two quotes brought back by one NBC News crew:

> "I think if you change your life, they're winning," says Captain Frank Carver. "So the more we continue our daily routine, better off we all are."
>
> At Pat's Country Bakery nearby, Joann Charters concedes she's still apprehensive. "It's a really scary feeling with kids in school. You don't know what's gonna happen," says Charters.

To accurately summarize these quotes you'd have to say: "Some people feel one way and some other people feel another way. Back to you, Tricia."

Joann Charters citing that it's scary because "you don't know what's going to happen" is right on. That's why it's scary: because you don't know what's going to happen—not because you do know, not because danger is advancing toward you, but because it is not.

TV news stories like this are filler, background, static, irrelevant. You don't need a reporter and a video crew to bring you man-in-the-street opinions. There are men on your street you can get opinions from. Or you could just talk to your friends and family.

WARNING SIGNS …

Any list of warning signs implies great risk. I recall a rash of reports about car-jacking in Los Angeles, and this list of warning signs:

Armed stranger approaches car;
Taps on closed window;
Looks around suspiciously.

And then they offered the checklist of precautions, given by an "expert on car-jacking." (Is there a college course on that?) The checklist:

Keep doors locked;
Don't let strangers into your car;
Drive away.

This is tantamount to:

"NEXT UP: CRIMINALS WHO HIDE OUT IN YOUR PURSE AND ROB YOU WHEN YOU GET HOME!"

Warning Signs:

Purse feels extra heavy;
Strange noises coming from purse.

OFFICIALS ADMIT …

"Officials admit that the incident could have developed into a full-fledged riot …" In this context, admit means that when a reporter asked, "If police had never reached the scene, and if a hundred other factors had fallen into place in an extraordinarily unlikely way, couldn't this have developed into a full-scale riot?" Yes, it could have—an admission.

EXPERTS …

It may seem you are getting expert advice on the news, but that's far from so. The moment you edit what an expert says, it's just words you might as well put in the blender. Would you let a TV news crew mediate your doctor's advice? Imagine being challenged by a difficult illness and your doctor's compassionate and complete 30-minute presentation was edited down to 23 seconds.

That's what the local news brings you: expert opinion edited, mediated, and minimized by non-experts who ask questions designed to elicit the most alarming responses. "Yes, yes, Dr. Stevens, but if it did happen, it would be terrible wouldn't it?"

NAMES

MONIKERS

When the news media assign a nickname to a wanted criminal (e.g., The Night-stalker, The Hillside Strangler) or to a disease (Legionaire's or Flesh-Eating Diseases), it is indicative of a hoped-for series of reports. When it's a type of crime (Follow-home Robberies), a trend is not far behind.

For example, freeway shootings and "Road Rage" led to all these headlines: "AGGRESSIVE DRIVERS TURN FREEWAYS INTO FREE-FOR-ALLS." "ROAD RAGE: DRIVEN TO DESTRUCTION," "HIGHWAY VIOLENCE SPREADING LIKE AN EPIDEMIC."

Next comes "Officials are concerned," and soon enough—as with Road Rage, you've got hearings before the House Subcommittee on Surface Transportation, and somebody (in this case, committee staff member Jeff Nelligan) calling the issue, "A national disaster." Presumably, Mr. Nelligan would tone that down a bit today—all of us having found a new meaning for the words "national disaster."

I TOLD YOU SO ...

An NBC News story quotes a member of a university task force on weapons of mass destruction: "We've been talking about this for years and people in general have not been interested." Is there some surprise there—that someone on a task force about weapons would be talking about weapons? The intended implication of these stories is that if someone had just listened, this could all have been prevented. How could discussions at some college task force have been used to prevent anthrax scares? If we had listened, what would be different? This is like an earthquake happening and earthquake experts saying, "We warned you." Yes, you did; you said there'd be an earthquake sometime. If only we'd listened.

DISASTER UNREADINESS ...

These are stories where TV news people cannot lose. They ask hospitals or public health officials or the utility company or the fire department if they can handle a disaster of X magnitude. If the response is yes, they just keep upping the disaster magnitude until the response is no.

Here's an example from NBC News: "A survey of 30 hospitals in four states and Washington, D.C., found them ill-equipped to handle a widespread biological disaster." A guaranteed fear-inducer, pokes right at our insecurity. First off, just asking the question implies that a "widespread disaster" is coming, and it's even better if the survey was part of a "new study," because that implies that the question itself is well founded.

Either way, the basic premise of the story is true: If hospitals currently able to handle 500 patients an hour get 5000 patients in some terrible hour, they will be unprepared. The standard of care will drop. Is there something surprising about that? Do TV news writers think Americans assume there is some extra team of 200 doctors and an extra 5000 fully-equipped hospital beds waiting in their community somewhere just out of sight?

Indeed, hospitals are unprepared for that which they have never had to be prepared. Being able to deal with what predictably comes down the pike and putting your resources where they are most likely to be needed is good planning. An emergency room would have to trade some daily-used resource to be ready for mass casualties that don't appear to be coming. Yes, as the world changes and events change, so does preparation—but expecting hospitals to be fully prepared, for example, to treat thousands of inhalation anthrax casualties when there's been a few lethal cases in 30 years would constitute bad planning.

One can make an "unprepared" story about anything; America's police are unprepared for a "widespread crime disaster"; our supermarkets are unprepared for a "widespread food shortage." It all depends upon how you define the word widespread. Put a microphone in some official's face and ask if he's adequately prepared for an attack on the harbor by Godzilla, and you've got an unreadiness story.

WILL NEVER BE FORGOTTEN ...

"Being stuck in the elevator for six days is an experience Betty Hamilton will never forget." This is used as a measure of how serious an incident it was, but did anyone imagine she was going to forget it? "I think I was stuck in an elevator for six days, but I can't quite remember."

THE WRAP-UP ...

Pay attention to the very last line in news reports. They are rarely summaries, but rather are designed to keep the story open for more reports. Most often, the closing line takes a last bite at the fear apple, one final effort to add uncertainty and worry. "Many here are left wondering if it will ever be safe." "Fear continues its tight grip on this tiny community." "Whether more will die remains to be seen." In the world of TV news, frightening stories never end. We never hear the words "And that's that."

Let's put a few of these newsroom strategies together into a story and see how it looks. As the basis for our mock TV news report, I'll draw on something that actually happened to my assistant. Earlier this year, her wrist was injured when a dog bit her.

THE TEASER:

"NEXT UP: DOGBITES! THE BONE-CRUSHING POWER OF DOGS." Experts warn that even friendly dogs can bite, sometimes without provocation. And they're everywhere. A new government study estimates as many as 300 dogs per square mile, with the numbers climbing each year. How many backyards in your neighborhood are hiding a deadly menace? We'll tell you what experts say—when we come back.

THE STORY:

A shocking bite from the dog everyone described as "a little angel" leaves one area woman nursing her wounds. Dog-jaw experts say that even a small dog can produce as much as 500 pounds of biting force, and given the rate at which dogs breed, it's just a matter of time before more people are placed at risk. A former employee with the Department of Health says hospitals are unprepared for a major increase in dogbites, and officials are closely monitoring this situation that could pose a deadly threat to our nation's neighborhoods. Disturbing questions have been raised about loopholes in the licensing system, and observers point out that dogs who bite can receive licenses and be released into neighborhoods.

THE WRAP UP:

It's no surprise that many local residents are living in fear: "You never know when somebody is walking their dog right behind you. We're scared." Officials say links between the recent dog bite and one that occurred in the tiny town of Ames, Iowa have not been confirmed, but either way, it's a nightmare few will ever forget. And one that many fear will not be over in the morning.

Summary

A free and independent press is a cornerstone of a democratic republic. The duty is to report the news impartially with no slant or twist favoring any one viewpoint over another. Unfortunately, the foibles of humanity have generally corrupted that ideal with "fake news," "alternative facts," opinions, and political preferences. What is also influential in news reporting is the number of people watching (and being influenced), thus creating ratings, translating into more money from sponsors. The more fear and violence they report, the more viewers they attract. The use of subtle words and phrases is a powerful tool in getting and keeping faithful watchers.

Review Questions

1. Is fear of violence totally justified in society?
2. What makes fear a good motivator?
3. In what ways do the regular news outlets and social media generate fear when reporting violence?
4. Is it morally right for news media to "tease" viewers into watching by using subtle, and innocent sounding, words and phrases?

Note

1 www.psychologytoday.com/blog/two-takes-depression/201106/if-it-bleeds-it-leads-understanding-fear-based-media. Retrieved June 28, 2019.

2 https://gdba.com/media-fear-tactics/. Retrieved October 17, 2019.

12

MEAN STREETS AND SHOOTINGS

When a government is established, the first thing it tries to do is to monopolize violence. Its soldiers and police are the only ones legally authorized to exert violence against its citizens. The late President of France, Charles de Gaulle, said "The law must have force on its side"; behind every law is the barrel of a gun.[1] Limitations (in most nations) are placed on the degree or level of force police are allowed to manifest. The cardinal rule is "the only amount reasonable and necessary" if the officer's life or the life of another is in immediate danger. "Experts say officer-involved shootings are all about an officer's perception of danger, which can be a tricky thing to gauge. What one officer might perceive as dangerous, another officer may not."[2]

Police Shootings

In 2016, law enforcement agencies in San Diego County (California) announced a 20-year study of 358 police shootings. Among the findings were two outstanding points: (1) the average age of the officer involved in the shooting was 35 and (2) the average length of employment was 10 years.[3] These findings are compatible with the timeline of police attitudes found in Chapter 14. Approaching the 10-year mark the officer has become cynical and distrusting, quick to use excessive and unnecessary force. In the past, experts have said that reevaluating law enforcement tactics—specifically placing less emphasis on force and more on face-to-face interactions (commonly referred to as "community policing")—could help cut the shooting rates, at least in the short term. A more complicated issue is the relative lack of trust in police officers. At least 1,058 people were killed by law enforcement officers in 2016 and 996 in 2018.

The number of police officers killed while on duty has risen faster than years past. An analysis by the National Law Enforcement Officers Memorial Fund showed that 135 officers died on the job in 2016, the highest in five years but still well within the recent yearly average. For the past 10 years, a yearly average of 151 officers have died on the job, according to the memorial fund. "We must never forget that 900,000 law enforcement officers nationwide risk their lives every day for our safety and protection," said the memorial fund's president and CEO, Craig Floyd, in a statement. "This year (2017), 135 of those men and women did not make it home to their families at the end of their shift."[4]

> More law enforcement officers—44—died in the line of duty in the United States in 2018 than in 2017, reversing a one-year decline. … Of those killed in 2018, 134 were male and 10 were female. The average age was 41, with an average length of 12 years of service. … Most of the officers were shot or fatally injured in car crashes. Others died from heart attacks, drownings, cancer and other illnesses developed by responders to the World Trade Center attack.
>
> The 52 firearms related deaths were a 13 percent increase over 2017. Fourteen of those occurred while the officer was trying to place an individual under arrest. … Nearly two-thirds were shot and killed by a handgun; four officers were fatally shot by their own weapons after being disarmed.[5]
>
> "Whenever a police officer is involved in a fight, there is always a firearm present … that is the one belonging to the officer."
>
> *A criminal justice student/police officer*

"We're 20 weeks into 2019, and so far 19 officers have been shot and killed in the line of duty across the country. That's a decline from last year. Firearm-related deaths for officers are about a third fewer than what they were at this time in 2018," according to the National Law Enforcement Officers Memorial Fund. But to those affected, it's little comfort. "Even one death is too many," NLEOMF spokesman Steve Groeninger told CNN. "Every number represents a shattered family, a shattered department, and a community that is grieving."[6]

Law enforcement officers face great fear-inducing and life-threatening intimidation of violence while on duty. This places in them a mindset of the possibility of serious injury, even death, forcing them to be on edge, always on the alert for indicators of impending doom. Thoughts of death and fear are pushed to the back of the officer's mind, causing strain and stress, so that when a fearful event presents itself, the individual officer may respond with an uncontrollable emotion, leading to lethal violence, which in the eyes of a bystander will seem unreasonable and unnecessary.

> Liberal politicians view Criminal vs. Law Enforcement Officer as some forms of Medieval Jousting or Marquees of Queensberry boxing with rules. It's not a fair fight! Never suppose to be a "Fair Fight." Officers are supposed to win every time! The fairness (legality) is to be determined in a Court of Law at a later time, not in the streets or offices of politicians jockeying for votes. End of conversation.
>
> *Retired Police Officer*

The Supreme Court Weighs in About Use of Lethal Force

The U.S Supreme Court ruled in *Tennessee V. Garner*[7] that lethal force may be used if it is necessary and the police have *probable cause* to believe the individual poses a significant death threat. *Graham v. Connor* created the "Reasonable Officer Standard," which dictates that an officer with the same training and experience, in the same circumstances, at the moment force is used would do the same thing. This is without using 20/20 hindsight with the facts known to the officer at the time of the incident.

Justice White delivered the opinion of the Court:

> This case requires us to determine the constitutionality of the use of deadly force to prevent the escape of an apparently unarmed suspected felon. We conclude that such force may not be used unless it is necessary to prevent the escape and the officer has

probable cause to believe that the suspect poses a significant threat of death or serious physical injury to the officer or others.

Probable cause is most often defined as "those facts and circumstances that would lead a reasonable person to believe ..." A police officer is considered a "reasonable person" and he must be able to "articulate the totality of circumstances" that led him to believe his life or the life of another was endangered.

Typically, the use of force generally falls into one of three categories:

- Appropriate use of force—the required force necessary to effect a legitimate arrest or protect public safety.
- Biased and prejudicial-based use of force—this often comes from profiling, stereotyping, and other biased-based interaction. Many times it is the result of disconnected interaction within the communities police officers serve. The officers view all interactees as combatants, not citizens to be protected and served. Force is used to drive compliance, respect or fear produces an overseer mentality, develops self-entitlement by the officers, and destroys public trust particularly in minority communities.

Walter Scott, the unarmed man who ran away from South Carolina police officer Michael Slager during a minor traffic stop, was of no threat to the officer or anyone else. He was stopped for a brake light in his rear window that was not working. While he may have had outstanding warrants for child support neither the public nor the officer were at risk and he did not deserve to be shot in the back as he was running away.

- Tactical errors—this is by far where we often see the highest level of questionable deaths by police officers, either through a lack of training, complacency, or failure to follow department protocols and operating procedures. In these situations, the officers find themselves behind the reaction curve and are forced to do recovery.

The case of Tamir Rice (the 12-year-old boy victim of a police shooting) in Ohio is an example—65 percent of officers are often killed inside of 10 feet from their attacker. In this case the officers responding to a police dispatch call "of a male black sitting on a swing and pointing a gun at people" in a city park, pulled their vehicle right next to Rice. Still in the car, they told Rice to drop the gun. Within two seconds the officers shot and killed Rice. After the officer fired upon the kid he retreated toward the rear of the vehicle, establishing both distance and cover/concealment.

There were several questions about the police tactics. Why didn't he start from that cover/concealment position initially? Why wasn't the initial response to the young man to stop and put his hands in the air given from the vehicle's PA? Why not maintain distance from out of the car, behind cover, get compliance. (Young man, keep your hands over your head. I need you to drop down to your knees and put your hands behind your back or lay on your stomach.)

Another example of police hasty over-reaction and poor judgment is the case of Levar Edward Jones who was shot by South Carolina State Trooper Sean Groubert during a routine traffic stop. The dashcam video shows the officer stopping Jones at a gas station/convenience store. He asks the man for his license; the man reaches back into the car for his license and the officer shoots him multiple times.

It is not the public's job to understand how to interact with law enforcement. Officers must be specific in their instructions. In neither of these situations did the officers wake up that morning with the desire to kill or injure a law-abiding citizen. The mistakes they made were born of poor tactics and very possibly fear for their lives. Fear overrode good judgment and training. When our actions are driven by fear vs. the presence of danger, they are often inappropriate, excessive, or unjustified. We must learn to separate *fear* from *danger*. Another man or woman is not responsible for your fear, though they may be responsible for the danger to you. Fear is personal.

Facing a Gun

Responding to a burglar alarm, an armed robbery in progress, or making a traffic stop, being called to the scene of a domestic dispute, or any one of an endless number of circumstances requiring a police presence, an officer never knows what will trigger an armed confrontation. He may practice hours upon hours at a firing range, shooting at paper targets that don't shoot back. He is in a safe environment, among friends and co-workers and the procedure becomes routine. There is no pressure except to meet a minimum qualifying score. But shooting at an individual who is shooting back at you or holds a gun is another issue. No matter how much time on the range gaining accuracy and practicing various scenarios ever prepares an officer for that one defining life or death situation. Maybe in his entire career he has no occasion to use his weapon or ever confront one in a potentially lethal way, until … that day.

One officer wounded in a gun fight said, "It was frightening, very scary, all the words for that. But you don't realize that at the time … the scariness comes later when you sit back and critique yourself. Then you get scared."

In the eyes of one convicted police killer:

> Most police officers are not conditioned for a real gun battle. They practice on unmoving paper targets, they're overweight, burdened down with heavy body armor and equipment, they drink a lot of beer, they stay up at night and they don't sleep very well. I think most have got ulcers, heart problems, premature illnesses, his mind may be disturbed by family problems, he is nervous and ratcheted up with emotion. … I think most people who come from high crime areas, they view the police as the enemy, a threat. Obviously, a person who carries an illegal concealed weapon, he's not ready to submit because he has the audacity to possess a firearm in the first place. Therefore, it's a matter of who gets the drop on who, the police don't know when a gun will be pulled but he can't do anything until he sees a movement that makes him think his life is in immediate danger. I know what I am going to do and when. The advantage is mine.[8]
>
> "There's little resemblance between what we train officers for and what they actually encounter on the street," … "There are glaring deficiencies in the way cops are prepared for what turn out to be fairly typical circumstances in gunfights." … when only one officer fired during an encounter, the average hit ratio was 51 percent. When an additional officer got involved in shooting, hits dropped dramatically, to 23 percent. With more than 2 officers shooting, the average hit ratio was only 9 percent—"a whopping 82 percent declination."[9]

In a social media posting, Kymlee Murphy, VP of Operations and Firearms & Tactics Instructor for Commonwealth Criminal Justice Academy in Alexandria, Virginia, said: "There are things called blood lessons. Situations that happened to many officers not wanting to harm another individual who used a furtive movement to retrieve a hidden weapon and kill the officer." Research South Carolina Trooper—Cpl. Mark Coates (see YouTube murder of Trooper Mark Coates). Also watch the Kyle Dinkheller video (YouTube).

> The long and short of it is, it's horrible to have to be put in a position in which you have to correct someone's behavior by using force in the first place. Law enforcement officials have to make split-second decisions in tense and rapidly evolving circumstances, based on their training and experience as well as the above mentioned criteria for reasonableness.
>
> Let me leave you with this, I have been in a position and some of my loved ones are in a position where they must enforce the law—because without consequences people do whatever the hell they want to each other ... But at the end of the day—We are going home to our families. If your actions are going to infringe upon our rights to go home to our families, because you want to get away with crime, or other intent that we cannot discern, then the consequences are on you.

The following paragraph[10] describes the events of the night that Deputy Dinkheller was murdered.

> Deputy Kyle Dinkheller, Laurens County, GA, was minutes from being off duty when he encountered a speeding pickup truck going 98 mph. The deputy was an ICE (Interstate Criminal Enforcement) officer that dealt with traffic infractions, speeding and the occasional drug bust. This was a low risk or unknown risk stop for speeding. He radioed in the speeding infraction, made a U-turn in the median and pursued the vehicle. The driver, Andrew Brannan, stopped his vehicle, exited and started a crazy, dancing jig in the middle of the road while swearing at the officer and shouting "I'm a god-damned Vietnam vet." At first, he ignored Dinkheller's commands to step towards the deputy, which always began with "Sir." When he finally complied, he attacked the deputy and a scuffle ensued. The deputy implemented the use of his asp and ordered Brannan to "get back." This procedure was repeated, but after what appeared to be a second scuffle, the suspect returned to his vehicle and retrieved a M-I Carbine from under the seat. The first shots were fired nearly 50 seconds after Brannan returned to his vehicle despite the deputy's commands. Brannan ignored the repeated commands to put the gun down and Deputy Dinkheller apparently fired the first shot. Brannan, a Vietnam veteran, advanced firing on the deputy. Dinkheller returned fire, but succeeded only in breaking a window in the driver's side of the pickup and wounding Brannan in the stomach. Using "suppressive fire," Brannan systematically, methodically shot Dinkheller in the arms, legs, exposed areas that would not be covered had Dinkheller been wearing a bulletproof vest, slowly executing him. Reloading his weapon Brannan continued firing with the final death shot to Dinkheller's right eye. "The entire incident was videotaped by a camera in Deputy Dinkheller's patrol car. On January 28, 2000, the suspect was found guilty of murder and was sentenced to death two days later."

Continuing with the conversation with Kymlee Murphy:

> Fear will always have a place (minimal hopefully) because it keeps people safe and links with intuition. However, overcoming that fear in tough situations is even more

> important. No matter the dilemma you have to persevere. Lean on your strength, and trust that you know the answer. Because if others are to follow you, your compass should already be sound. Also, taking the time to reflect and soul search is important for growth. Proving to yourself that you can comprehend where you are weak, and challenging yourself to become strong is providing yourself with an internal solution.

One police training officer, a survivor of a shooting incident, gave solid advice in saying: "When you think the guy holding the gun is 'targeting' you … move! Fall down, roll, jump to the side or whatever you can do to change the shooter's perspective." Very few shooters are proficient and trained in accurate shooting, especially at a moving target. Studies have concluded that the odds of being hit while moving or running away are approximately 4 in 100.

Shooting Contagion

"Rat a tat, tat, tat, tat" or "Get 'em off, get 'em off, off, off, off" (said to the tune of the *William Tell Overture*), when referring to contagious shooting or firing as many rounds as possible in a limited time frame: This is often the refrain heard at a police training firing range.

"'Contagious shooting' refers to the phenomenon in which multiple officers will start shooting once one officer has opened fire—even if no real threat is present. Experts say the reaction is a response to sudden fear."[11] Dr. Carole Lieberman, a Beverly Hills forensic psychiatrist and assistant clinical professor at UCLA's Neuropsychiatric Institute in Los Angeles, California said, "When a fellow officer starts shooting it sends off an alarm, warning the others that they must be in lethal danger. … Not only is there ostensibly a 'killer on the loose,' but they fear they can no longer count on their fellow officers to protect them. … The officer who starts the shooting is responsible for where he aims and how many times he shoots at the alleged 'criminal,'" she says. "The other officers who become affected by the 'contagion' are responsible for panicking, becoming blinded by the outbursts of gunfire, and not restraining themselves enough to assess the situation."[12]

"Shootings involving only one officer, an average of 3.59 police rounds were fired. When 2 officers got involved, the average jumped to 4.98 rounds and with 3 officers or more to 6.48. The number of rounds fired per officer increases in multiple-officer shootings by as much as 45 percent over single-officer shootings."[13]

In one contagious shooting incident in Los Angeles, sheriff's deputies fired in excess of 50 rounds at a car driven by a drunk driver who had struck a deputy's car following a high speed pursuit. One deputy actually reloaded and emptied his weapon twice. In another incident a year prior, deputies were chasing a driver who apparently was under the influence of drugs and refused to stop. In approximately 18 seconds, deputies fired over 120 rounds at the driver's pickup truck. Four rounds hit the driver (none fatally), one hit a deputy sheriff, eleven hit the sheriff's vehicles. Eleven hit five homes in the area and the rest were unaccounted for.

"Hey! Show me your hands! Stop! Stop!" an officer yells. Then one of the officers begins shouting, "Gun, gun, gun!" Gunfire erupts in a quiet night-time neighborhood in Sacramento. In a matter of seconds, 20 shots are fired and Stephon Clark lies dead. Police were responding to a vandalism call reporting that someone was breaking windows in cars on the street. An independent pathologist conducted an autopsy, finding that Clark had been hit eight times, seven in the back. No gun was found, only an iPhone Clark had in his hand. Perhaps officers truly believed Clark was armed with a gun; it is very difficult to distinguish in limited lighting. The biggest question in this scenario remains: Why were 20 shots fired? The simple answer is: contagion shooting.

Contagion shooting may be a Pavlovian response from training. Several officers at a time train by forming a "firing line" and follow commands such as "Ready on the right, ready on the left, begin firing!" All the officers respond to the command and begin firing at once. In instances of timed shooting practice, when the targets are presented, one officer will fire quicker than the others, becoming the "go ahead" for all the others to begin shooting.

A former chief of police in the Midwest who had also worked as a patrol officer in Harlem said, "Four times I came close to pulling the trigger and thank God, I didn't. It's only two and a half pounds of pressure. You can't undo that decision. When the blood is pumping, in the heat of the adrenaline, you've got to maintain your self-discipline."

Militarization of Police

> State and local police departments are increasingly accepting the military as a model for their behavior and outlook. The sharing of training and technology is producing a shared mindset. The problem is that the mindset of the soldier is simply not appropriate for the civilian police officer. Police officers confront not an "enemy" but individuals who are protected by the Bill of Rights. Confusing the police function with the military function can lead to dangerous and unintended consequences—such as unnecessary shootings and killings. What is clear—and disquieting—is that the lines that have traditionally separated the military mission from the police mission are getting badly blurred. The job of a police officer is to keep the peace, but not by just any means. Police officers are expected to apprehend suspected law breakers while adhering to constitutional procedures. They are expected to use minimum force and to deliver suspects to a court of law. The soldier, on the other hand, is an instrument of war. In boot camp, recruits are trained to inflict maximum damage on enemy personnel. Confusing the police function with the military function can have dangerous consequences. (A large city police chief) has noted, "If [cops] have a mindset that the goal is to take out a citizen, it will happen.' … Thus, the use of paramilitary tactics in everyday police work should alarm people of goodwill from across the political spectrum.[14]

Police as Warriors

> A warrior is a person of courage and skill in battles and conflicts; Warriors are those who choose to stand between their enemy and all they love or hold sacred; A warrior is the protector of his family and those in need; A warrior is the guardian of the ways of honor and courage; A warrior will not put his own desires above what is right; Above all, a warrior is always motivated by his utmost duty.[15]

In the words of the famous Sioux leader Sitting Bull:

> Warriors are not what you think of as warriors. The warrior is not someone who fights, because no one has the right to take another life. The warrior, for us, is one who sacrifices himself for the good of others. His task is to take care of the elderly, the defenseless, those who cannot provide for themselves, and above all, the children, the future of humanity.[16]

All of these definitions could easily be ascribed to military personnel and, of course, police officers.

"Record numbers of law enforcement officers are being killed and in order to survive, a certain mindset must be maintained."[17] Police have long considered themselves warriors of peace. They train to develop a style of passive-aggressiveness; showing calmness but full of tactical awareness, always suspicious, cynical, and alert for the possibility of that unexpected instant they must be in full adrenalin combat mode.

Following, is a heavily edited and revised article discussing the *Path of the Sacred Warrior* with the words *police* and *officer* substituted for *Sacred Warrior.* [18]

> Our life's journey begins with the awareness that we are mortal beings, that we are going to die, the only things we don't know are when, where and how. The police officer walks his path with Death at his side. A police officer/warrior fears death, dares death and respects death. [Officers] live with death at their side, and from the knowledge that death is with them, they draw the courage to face anything. The worst that can happen is that we have to die, and since that is already our unalterable fate, we are free; those who have lost everything no longer have anything to fear.

Background stories of police officers may reveal a childhood filled with inner turmoil and outward aggressiveness, usually to defend another child who is being harassed, hazed, and bullied. They are keen to explore the world and they don't want anyone or anything to get in their way. They may fight with their siblings or test the parents mercilessly. They often seem to have come into life with an excess of energy, involved in sports and interested in physical activities. Their temperaments are fiery; their wills, strong.

They learn from the proximity and potential of Death that they need to develop humility, patience, and an ability to keep a clear head—or, at least, to clear one's head, fast! Yet they also learn that this closeness to death is a mirror to their own fears, and as such deserves respect and, even, love. From this realization, they go on to learn self-defense and self-reliance. Obviously, this is a path of courage. The more courage one shows, the more professionally they are respected.

What is bravery or courage, anyway? This question is central to the officer's maturation as an officer. It does not involve contempt. It is wasteful to feel contempt for people or other entities. A Native American warrior speaking to a group of White Americans put it this way:

> You people have such anger and fear and contempt for your so-called criminals that your crime rate goes up and up. Your society has a high crime rate because it is in a perfect position to receive crime. You should be working with these people, not in opposition to them. The idea is to have contempt for crime, not for people. It's more useful to think of every individual as another you—to think of every individual as a representative of the universe. Even the worst criminal in life imprisonment sitting in his cell—the center of him is the same seed, the seed of the whole creation.[19]

Detachment is important. An officer needs to be free of excessive burdens. He needs to be free to think clearly, and move at a moment's notice. He also needs to be able to live in the present. In order to cultivate detachment, an officer must develop his sense of humor and a great sense of resourcefulness. These become his shields. He can feel strong and passionate emotions and then let them pass through him. He can laugh at himself.

But there is a danger in detachment. A police officer can become so self-reliant that he becomes arrogant and uncompromising. He becomes incapable of compassion. He covers all his fears with braggadocio and his fists, baton, and weapon.

Good officers have reputations for great hearts, and even the strongest, most skilled, most dangerous warrior becomes sacred when he puts himself in service as a Guardian or a Champion to a child, a needy group, a holy place, a worthy task. Most of all, the police officer is at the service of those who truly require him. *He does this for them, because his self-imposed belief is "I'm here to help."* His love and services are free, without attachment or expectation—unconditional. He knows, perhaps more than anyone else, that to truly love is the most dangerous and most daring act a police warrior can perform.

While they are much admired and honored for law enforcement skills, police officers should be even more respected and loved for their compassion and kindness. In the past, Sacred Warriors battled for the protection and survival of their tribes, and for personal satisfaction. This is still true, but in our Age, the definition of "tribe" can vary. The Sacred Warrior who travels on "A path with a heart" must find her own sacred battlefield. The fight may be for justice, or peace, or respect—whether personally or publicly.

An American/Eskimo quote:

> Great are the tasks ahead, terrifying are the mountains of ignorance and hate and prejudice, but the Warriors of the Rainbow shall rise as on the wings of the eagle to surmount all difficulties. They will be happy to find that there are now millions of people all over the earth ready and eager to rise and join them in conquering all barriers that bar the way to a new and glorious world! We have had enough now of talk. Let there be deeds.

> *The problem with many of the "Warriors" is that there is no off switch and that they march to their own mission statement and not that of the department for which they work for. Sure, tactically, they are the ones you want but when it comes to the things that they deem not important, then it is "Too hip, gotta go." Their little problems ultimately become big ones.*
>
> *David Gemmell,* Legend[20]

> *Heroes are people who face down their fears. It is that simple. A child afraid of the dark who one day blows out the candle; a women terrified of the pain of childbirth who says, "It is time to become a mother." Heroism does not always live on the battlefield. … There's no shame in fear. But understand this—the coward is ruled by fear, while the hero rides it like a wild stallion.*
>
> *David Gemmell,* Dark Moon[21]

Russian religious and political philosopher, Nicolai Berdyaev (1874–1948), described the differences between modern civilized Man and ancient warriors:

> *Modern civilization cannot endure cruelty, pain, and suffering and is more merciful than men of the past. But this is not because he is morally and spiritually higher than they. He fears pain and suffering more than they did; he is more effeminate, less firm, patient and courageous than they. In other words, he is spiritually less strong.*

> *You wake up every morning to fight the same demons that left you so tired the night before.*
>
> *Anonymous*

Summary

Fear is an insidious devil that rides in the patrol car with an officer. Danger lurks behind every contact. It often seems officers are killed at an alarming rate. Analytics prove the majority of police killed by gun fire in the line of duty is the result of the officer's mindset, mistakes, lack of training, violation of departmental policy, or poor judgment.

Contagion shooting is a Pavlovian response to training where officers shoot on command and all begin at once. In the field, one shot by one officer may set off a volley by other officers.

Many officers adopt a belief that they are warriors protecting a helpless citizenry but also often develop a detachment from their own feelings. But there is a danger in detachment. They can become arrogant and uncompromising. They can become incapable of compassion. They cover all their fears with braggadocio and fists, baton, and weapon.

Review Questions

1. Explain *Tennessee V. Garner.*
2. In the named police shootings illustrated in the text (Jones and Rice), did the officers act properly?
3. Is police militarization a beneficial trend or is it harmful to police–community relations?
4. There are several definitions of a warrior mindset. Do you endorse or reject that description for a police officer?
5. List and describe the mistakes made by Deputy Kyle Dinkheller on the night he was killed. In your opinion what should he have done to avoid being killed?
6. The common belief by many citizens of society is that police sometimes over-react to violence by using unnecessary and unreasonable force. What do you believe motivates that reaction by the police? Cite examples.

Notes

1 Toffler, Alvin; *Powershift*, Bantam Books, NY; 1990; p.38.
2 Danylko, Ryllie; "Examining Police Training: How Officers Are Taught to Deal with Armed Suspects"; www.cleveland.com/metro/index.ssf/2014/12/how_police_are_trained_to_deal.html; December 8, 2014; Retrieved April 25, 2018.
3 Zabala, Liberty and Stickney, R.; *Officer-Involved Shootings Study Released: The study looked at 358 officer-involved shootings that occurred in the county between 1993 and 2012.* www.nbcsandiego.com/news/local/Officer-Involved-Shooting-San-Diego-County-District-Attorney-292716271.html. Retrieved April 27, 2018.
4 McCarthy, Ciara; "More US Police Officers Killed In 2016 – But Number Still Below 10-Year Average"; Thursday 29 December 2016, 12.57 EST. www.theguardian.com/us-news/2016/dec/29/police-killed-2016-average. Retrieved April 17, 2017.
5 Stanglin, Doug; "144 Police Officers Died in the Line of Duty in 2018, Reversing a One-Year Decline"; *USA Today;* January 1, 2019; www.usatoday.com/story/news/2018/12/27/police-deaths-144-killed-line-duty-2018/2423797002/. Retrieved January 6, 2019.
6 "In the First 20 Weeks of the Year, 19 Police Officers Have Been Shot to Death in the Line of Duty"; posted 2:58 pm, May 20, 2019, by CNN Wire; https://wreg.com/2019/05/20/in-the-first-20-weeks-of-the-year-19-police-officers-have-been-shot-to-death-in-the-line-of-duty/. Retrieved July 8, 2019.
7 471 U.S. 1,105 S. Ct. 1694, 85 L. Ed. 2d 1,1985 U.S.
8 Adams, Ronald, J.,Thomas M. McTernan, Charles Remsburg. *Street Survival: Tactics for Armed Encounters*; Calibre Press, Northbrook, Illinois
9 *Study reveals important truths hidden in the details of officer-involved shootings;* www.policeone.com/officer-shootings/articles/117909-Study-reveals-important-truths-hidden-in-the-details-of-officer-involved-shootings/. Retrieved 04/26/18.
10 www.liveleak.com/view?i=969_1263249923. Retrieved April 28, 2018.
11 Baram, Marcus; "How Common Is Contagious Shooting?"; http://abcnews.go.com/US/LegalCenter/story?id=2681947&page=1; November 27, 2006; Retrieved April 13, 2017.
12 Ibid.
13 "Study Reveals Important Truths Hidden in the Details of Officer-Involved Shootings"; www.policeone.com/officer-shootings/articles/117909-Study-reveals-important-truths-hidden-in-the-details-of-officer-involved-shootings/. Retrieved April 26, 2018.
14 Cecilia, Diane; *Warrior Cops: The Ominous Growth of Paramilitarism in American Police Departments* CATO Institute Briefing Papers No. 50 August 26, 1999; Retrieved August 9, 2015.

15 Agnes Whistling Elk, from Lynn Andrews' book *Medicine Woman*, 1981, Harper & Row, NY. From the book *Daughters of the Earth* by Carolyn Niethammer, 1977, MacMillan, NY.
16 Motivational sayings shared on Facebook.
17 Katz, Marty; "Police Warriors: Servants of the Peace." July 28, 2011. www.policeone.com/police-jobs-and-careers/articles/3907091-police-warriors-servants-of-the-peace/. Retrieved September 7, 2015.
18 Andreas, Peggy; *Path of the Sacred Warrior*; First circulated on the newsgroup alt.religion.shamanism, this essay was published in *Towards 2012 Part I: Death/Rebirth* (The Unlimited Dream Company, 1995). Retrieved September 18, 2015 from http://dreamflesh.com/about/contributors/#peggy.
19 Boyd, Doug; Mad Bear, from *Rolling Thunder*; Dell Publishing Company, 1974.
20 www.goodreads.com/work/quotes/1805413-legend. Retrieved February 25, 2019.
21 www.goodreads.com/work/quotes/1907094-dark-moon. Retrieved February 25, 2019.

13

ANGER AND FEAR

Anger

Some sources have mentioned Anger as being the "lion of emotions," as it often lifts a depressed person to his baseline of normal activity. "Anger is the emotion that expresses dislike or opposition toward a person or thing that is considered the cause of aversion. Psychologists consider anger a natural emotion needed for survival."

In a "fight or flight" moment, anger can be the catalyst pushing the individual to fight. As adrenalin spills into the system, fear (flight) and anger (fight) are at odds to control the emotional moment. If anger becomes the controlling factor, fear is overridden and blind fury becomes lord over the moment, resulting in fight, which could be for survival. Fear can make a person bigger, stronger, faster. Anger begins with shortened breathing, a rise in body temperature, heart beat and blood pressure. It controls the muscles and nerves with a shutting down of the senses. The person is overcome by a white-hot flaming blinding fire and all he can see is "red." He becomes blind to everything except the target of his anger, wanting to strike out at it … to destroy it.

When out of control, anger can become destructive and will eventually lead to problems. Violence toward other people is one of the ways that people use to express their anger. Some individuals act on their violent thoughts whereas others do not. Why is that? It's all about self-control. To have self-control is to have a well-functioning life. Some people just have better control of their emotions and some just don't.

> Violence is not a product of mental illness; violence is a product of anger. When we cannot modulate anger, it will control our behavior. … Nor is violence generally the action of ordinary, stable individuals who suddenly "break" and commit crimes of passion. Violent crimes are committed by violent people, those who do not have the skills to manage their anger. Most homicides are committed by people with a history of violence. Murderers are rarely ordinary, law-abiding citizens, and they are also rarely mentally ill. Violence is a product of compromised anger management skills …. Violent crimes are committed by people who lack the skills to modulate anger, express it constructively, and move beyond it.[1]

Many Congressional Medal of Honor winners, exposed to a life-threatening situation, conquered fear by turning it into anger and rage so uncontrollable that they overcame overwhelming odds to kill and capture the enemy. When asked why they did such a courageous thing, the answers have been: "I was scared and angry." "For my friends, the brother on my right and left and I was angry they were being killed." "*Greater love hath no man than this, that a man lay down his life for his friends.*"[2]

"Anger can bring forth behavioral improvements (uplifting and suppressor of depression), however uncontrolled anger can cause social and personal problems." Rage and anger issues have been seen in professional athletes, especially players in the National Football League, as well as in police officers. Football players and police officers are naturally aggressive and have many similarities. An unstated element in the violence of a professional athlete and a police officer is the underlying factor of fear.

On the gridiron, a football player faces other players who are intent on hurting him on every play, creating suppressed fear. He counters his opponent by being as fierce, aggressive, and angry as he possibly can become. Often the anger is carried over to his private life and he goes home to pound out his anger (suppressed fear) on his wife.

Fear in a Violent World

Police work in a violent world. They often see the seamiest side of society and, of course, the result of violence as it leaves its impression on members of society. Every call and assignment a police officer receives holds potential danger and life-threatening elements in the unknown. Responding officers never know what they will be confronted with until arrival at a dispatched scene. Often the only prior information the officer has is what he or she is told by the dispatcher. In a high-profile case in Southern California, officers were dispatched to a convenience store to respond to "a robbery" with the suspect escaping on a bicycle (in California a robbery is defined as the taking of property of another through the use of force or fear). Being given that description as a "robbery," the officers arrived, thinking the robber was probably armed with a gun. This was the mindset of the officers as they drove to the scene with adrenalin pumping though their sympathetic nervous system.

Approaching the scene of the crime, officers saw two people on bicycles coming from the vicinity of the convenience store. Based on the officers' experience and the information given by the police dispatcher (who may have described the incident as it was reported by the victim—"I've been robbed!"), the officers drew their weapons upon exiting the car and stopped the young boys on the bicycles. As the officers were approaching the boys, a third boy ran to the bicyclists. Offices commanded him to raise his hands also. He momentarily complied but kept lowering and raising his hands, even extending them toward the officers with empty palms out. Several times the officers warned him to stop moving his hands or "you'll be shot." In a video from the officers' car's dash camera, it is apparent the officers were becoming angry that their commands were not being followed. The young man seemed to turn and, according to the officers, "he moved his hands toward his waist belt" and, being "afraid for their lives," began shooting. He was killed with eight shots to his body.

The tragic part of this incident is that the boys on the bicycle were friends of the "robbery" victim who had had his bicycle stolen and they were helping to look for it. As for the third boy, who was killed, he was unarmed and a friend of the other boys and was trying to explain to the officers. The police were later absolved of the killing in the interest of saving their own life because they were fearful they were endangered. Their defense was that they had faulty information; they believed the boy, by refusing to keep his hands raised and in sight, was going to pull a gun.

Unspoken during the aftermath of the incident was the adrenalin flow in the officers' bodies, which raised their "fight–fright" level to an extreme, causing them to take measures to protect themselves in the fear of being shot.

> Maybe it's stress or anger or adrenaline or disillusionment or a bullying nature or simple fear of getting killed themselves, but there is a problem if a cop cannot tell the difference between a menacing gangster and the far more common person they encounter whose life is a little frayed and messy.
>
> *David Horsey*[3]

In another southern California incident, officers received a dispatch to investigate a minor traffic accident. As they approached one of the drivers, the driver pulled a hand gun and shot the officers, one fatally, the other was hospitalized in critical condition. What the officers did not know was that the driver was recently released on probation from prison and an hour prior to the traffic accident, he had murdered another man.

Depression, anger, and fear can be correlated with fear of everyday living. The hidden fear of feelings of inadequately maintaining a strong economic base, fear of being an acceptable and loving parent and marriage partner, fear of loss of employment, and all the other common fears associated with adulthood. A teenager has fears of not being accepted, of being alone and friendless, of being bullied, of failing in school, not meeting parents' expectations and approval. As the fears (usually unfounded) compile, and depression begins taking its toll, an internal fury begins to grow until it reaches a dangerous level of anger resulting in an expressive blowout of violence. A bi-polar (manic-depressive) person can be dangerous to himself and others. In the depressive state, he considers suicide and means of harming himself. He may become explosively manic, as quickly as turning on an electrical switch, and burst into anger with vicious force.

In his 1948 dystopian novel, *Nineteen Eighty-Four*, George Orwell effectively described fear and anger as it can be felt in today's era:

> A hideous ecstasy of fear and vindictiveness, a desire to kill, to torture, to smash faces in with a sledge hammer, seemed to flow through the whole group of people like an electric current, turning one even against one's will into a grimacing, screaming lunatic. And yet the rage that one felt was an abstract, undirected emotion which could be switched from one object to another like the flame of a blowlamp.

Police Fear and Anger

Police officers are subjected to trauma-related incidents almost on a daily basis. Over time, the idealistic and friendly personality of the officer begins to change to one more hardened and angry. His anger is dispensed through intensity and forcefulness. It also becomes a way of compensating for fear of injury, perhaps impotence over the futility of dealing with traumatic incidents.

Psychologists divide anger into three categories. One type of anger is an instinctual reaction to being trapped or hurt (fear). Another type of reaction is to the perception of being intentionally harmed or mistreated by others. ("I was in fear for my life"). The third type of anger, which includes irritability, reflects an individual's personal character habits. An uncontrollably angry individual is susceptible to losing the capacity for making sound judgment and acting responsibility. Extreme anger is obviously self-destructive, as is anger that is not expressed externally and kept within. Anger is often misused by individuals who act angry as a means to manipulate others (causing others to react in fear).[4]

> Anger is a normal response to a traumatic event. Anger gives you the energy to act quickly and help yourself or others. Your body goes into a "survival" mode. After the event, when you no longer need to act, your anger goes away. Officers displaying acts of anger—often for no apparent reason or in reaction to a minor act of defiance or disrespect—may be suffering from or influenced by undiagnosed symptoms of PTSD. If you have PTSD (Post Traumatic Stress Disorder), your anger may get out of control. This means you lose your temper and may feel like harming others or yourself.
>
> When you have PTSD, you can get stuck in a "ready to act" mode. Your anger is always there, just under the surface. When something bothers you, you may not think about the situation before acting. You go into survival mode, and your anger flares up. If you are stuck in this mode, you may:

- Always be on alert. You may be quick to get angry, and look for situations where you have to be alert or where you could be hurt.
- Feel that anger is the best way to solve problems. You don't look for other ways, such as talking things over.
- Feel threatened and fearful about things that may not be dangerous."[5]

Institutional Fear

> It's not unexpected to find fear in an institution, regardless of the type of institution. A young person going to school the first time, going to junior high, then high school feels anxiety and fear until becoming adjusted to the environmental society. Adulthood means leaving the comfort and security of home as the person moves into college, the military or work. But those are, in reality, anxiety and stressful fears, totally unrelated to the fear invoked as a convicted (or suspected) law violator hears and feels the steel doors of jail and prison slam shut behind him. In addition to leaving behind the security of his normal life, now he must face daily threats to his very existence. Prisons and jails have a reputation of being "jungles" where only the strong survive. They may even be driven to a kill or be killed mental state. A prisoner never feels safe and fear is a constant reminder of the violence inherent in the closed society of gang members, killers, psychopaths and other miscreants of society. A fear of guards in a prison is also always on a prisoner's mind.[6]

Teenager in Prison

Following is a true story of a man, who, at the age of 19, found himself in state prison facing six life sentences and learning fear first-hand. (Thanks to Joseph Weedman for the following stories about being a teenager in prison.)

"In my teen years I had to handle many things from the time I was 16 by myself. Fights, ambushes and things of that nature already did not frighten me, and physical danger was only a sign of a challenge not imminent death. Needless to say, I was a tough kid beyond my years.

I have had firearms put directly in my face, even shot at. But the most frightening encounter I ever had was with the justice system and state prison. Daily threats in youthful offender camp is rough and real. Young guys all trying to make a name for themselves. It is actually a very interesting situation to find yourself in and if you are a skilled learner you can learn many life lessons. The weapons are very ingenious as are the levels of mentality and brutality; a 'lock and a sock,' razor blades moistened into bars of soap, shanks made from toothbrush ends, rebar metal fence meshing.

Three months into the arrest—I was just beaten by correction officers for passing Boston baked beans [candy] to a cell next door after a great meeting with my attorney. I was written up for reaching out of my cell in an open bay cell block; quickly taken out to be patted down for weapons. I was searched very roughly; they purposely hit me in my testicles both times they went up and down my legs. Long story short I ended up in lockdown.

A day or two later a young man came in and happened to be given the cell next to me. The trauma of what I was facing was already unfathomable. Two nights later blood began to trickle out of his cell. Coincidently count time happened to be at the same time. The guard doing the population count saw what was happening and called it in. Guards pulled this young man out of his cell after he had bled for a few minutes. He had slit his wrists. He was on the ground shaking while waiting for medical help and the guards stood around laughing and talking about how cold it must be on the ground. This is still only three months into my arrest and county jail stay. There is a level of fear that you feel in the treatment of someone who obviously may be mentally ill while lying there dying.

State Prison one year into arrest—An incident I recall very well was a severe beating to a guy who came from a prison in North Florida and belonged to the wrong gang. He became a target for a 'friendly hit.' He was beaten to within an inch if his life and happened to be the bunk next to me. I remember clearly looking at him and the clear flesh ripped from his lip. I was then told by guys from my city, as he laid there moaning from what was obviously death moans to 'cover him up' because it was count time. I don't think you know what fear is until you HAVE to partake in something you know is wrong. When inside all you want to do is get help.

I played cards with this guy my first two weeks in the compound. He already had fought seven guys back to back and won his first day. I built a kind of silent friendship with him and then I hear he's got a hit on his head and to keep my distance so I don't get caught in the mix. And then thirty guys I know are breaking broom sticks over him running over my bunk and me and jumping from bunks onto his head.

Scary? Fear? Those words are inadequate to describe the feelings and emotions. That heart wrenching pain and not being able to shake it. That comes from fear and was enough to break me.

I saw guys get shanked (knifed) at chow; heads crushed for not paying bets and being 'June Junes.' (A 'June June' is someone's property, or DO-BOY.) They take money and food from them in exchange for protection etc. It can be many things but the trade remains the same. 'You won't be messed with.' Extortion is a better word for it! It was constant fear inside. I say all the time 'I slept with one eye open.' One night I saw a kid get lit on fire in his sleep.

I would lose sleep, feel my heart coming out of my chest and my stomach become an empty pit of stomach aches, I had terrible dreams when I came home. Dreams of death that were very very real. I would wake up running, and screaming. I saw things in prison that toughened me, especially my mental stability with fear. But in that situation you don't always recognize your feeling as fear until after the event. I no longer suffer from the issues I had with fear. Though I still have anxiety I think my fear now is for my future.

Am I scared of death? I like to think I'm just scared of the unknown. So I make an effort to know as much as I can. I have slept with a night light my whole life."

Intimidated as he was, the prisoner speaking about some of the fears he saw and experienced learned to survive, but the memories remain and haunt him.

> "I remember what I have been through, the scary things I've seen, heard, and even some things I'm sure I was a part of. There was a time of my life, ages seventeen to nineteen, where many people thought I had a mental break down. I was not myself due to my situation. It was like a home cat gone stray and it changed my mentality and way of thinking for a little bit to better suit my way of life at the time. Something I may not even have fully recognized until recently. I think of that. Of it all! I am bigger than my FEARS. I still have a constant fear of death. Even though I do what I do for work. I'm also a firm believer it's what has kept me alive."

In a recent recounting of his days in jail, the young man said the following.

> "My 3rd encounter with Police Brutality. I was at North Broward County jail. At this location, there are open bay cells, with about 25–35 guys in each pod. I would receive commissary, so at times I'd give someone some Boston baked beans, or make a trade (this is illegal in jail, to exchange food). One day a kid in the cell next to me needed something to snack on. We always played cards together, and I figured what the hell, it's just 75 cents worth of candied beans. As I reached over to the cell next to me, to hand this kid the food, not even 2ft away, the C.O. comes on the loud speaker, 'Mr …, you're getting a DR' [disciplinary report]. This is how we end up in 30 day lock downs, with 23 hrs a day in a cell, and NO privileges.
>
> Little did I know, this was going to be a much bigger mess, than I ever could have anticipated.
>
> He calls me out to the hallway, and says 'you're getting a DR, and going to Lockdown.' Now at this rate, I don't care to argue for a few days in lockdown, so we proceed to the patdown phase. But, before this starts, two other officers were called to escort me. As one starts patting me down, he closed his hands around my ankle. Used to the procedure I was expecting a soft Pat up my leg.
>
> I was wrong.
>
> He came up with half his force, didn't slow, and then hit me square in the nuts. Naturally, I bend as my stomach clinched. I said to them 'Watch your hands you hit me in my nuts!' My hands still on the wall, they never left, he goes for my right leg, and this time repeats the attack. I start calling them names, and cursing, as I'm being put into cuffs. From there I was walked down the hall, but when we got to the stairs they decided to drag me up them, instead of allowing me to walk. When I got to the non violent lockdown cell, they laid me on the metal bunk in cuffs, belly down, and started to flip me over and over. Which led to one grabbing my leg and banging my knee on the bunk repeatedly. Then, they said 'go stand against the wall.' Well, I was not going to be raped or beaten today, by nobody, law enforcement included, and that position would make me more vulnerable, than I already was. Keep in mind, I was also already attacked.
>
> I walk toward the wall, and decide to get on one knee, allowing me the leverage I need to defend myself, if necessary. After some back and forth, they came forward and took the cuffs off.
>
> I was angry, and in pain. The second they turned to walk out, I got up and followed behind them, while I was calling them names. I was angry, and injured, with a scrapped up hip. About 2 steps in, 2 punches land square on my lip. I step backwards toward the

shelf that I can kneel under, for the coming attack by the officer behind, the first as well. As I sit there kneeled over I was kicked about 20 times, punched in the back of the head multiple times, and then grabbed and dragged downstairs. There I was thrown in a violent lockdown cell. The Sargent comes down to my cell, and throws me some paper towels. Told me 'if you don't clean up the blood, I will continue to beat your ass.' All the officers involved, had records of excessive force 2–6 pages long. And something that had been demonstrated to us on occasions with other inmates.

He then says 'you'll be charged with battery on a LEO [law enforcement officer]. And you will spend the rest of your time in 23/hour lockdown.' That led to a 7 month Admin 23/hr lockdown, on the 8th floor of the county jail, and the case that sealed my fate when it came to my freedom for the next 2 years.

Side note—I never threw a punch, or fought with my transport to lockdown. These things have been happening for a long time. It's just recent, that it's garnered attention from media."

Meme pas peur—"No Fear"

Have *meme pas peur* (no fear) as you go through the perilous night. The world is full of dangerous people and events. A simple trip to a service station/convenience store can turn into a life-threatening confrontation.

The same person as the narrator of the story above (Joseph Weedman), now a successful business and family man, shared his story of when he was a young man, recently out of prison, confronting two persons intent on robbing him and he showed no fear.

> In '08, about a week back into being in society [from prison], 2 young guys attempted to lock me in my local gas station, and rob me. Not only did I see it coming, because I talk the talk, I took it a step further. One stayed in the doorway, and locked it (click). And one made his way to the back, probably to see what I was purchasing.
>
> As one made his way to the back, I made my way to the front, not to be snuck from behind. I paid for my Heineken, and proceeded the five steps to the door. The kid in the door still has not moved out of the way, so I decided that I was going to walk right through him.
>
> (Bump), our chests meet, doors locked. He says to me, "run ya pockets." I take a step back, I pointed at his friend, and told them both to step outside so we can get on with this.
>
> The friend in the back, tells the guy at the door, "let's go then." When I get out there, I'm waiting patiently. But they make no move. So I say "What's good?" He responds, "Give me a ya shit." I respond, "we gon have to fight then."
>
> One of the dudes proceeds to take off his shoes, and his shirt. I'm standing there with my beer in my hand. I decided not to attack first while they were undressing.
>
> I decided to let them know "When you're done losing this fight, you will be walking home naked, and I need a pair of new shoes anyway, since I just jumped out the chain gang."
>
> These guys picked up their clothes, and said to me "We … we're just testing you."
>
> I responded with "That's how you get yourself killed."
>
> I watched them walk off.

From Assata Shakur's Autobiography, *Assata; An Autobiography*, published in 1987:[7]

> In prison it is not at all uncommon to find a prisoner hanged or burned to death in his cell. No matter the circumstances. These deaths are always ruled "suicides." They are usually Black inmates, considered to be a "threat to the orderly running of the prison." They are usually among the most politically aware and social conscious inmates in the prison.[8]

Fear through Intimidation

Niccolò Machiavelli (1469–1527), in his classic masterpiece, *The Prince*; Chapter XVII; "Concerning Cruelty and Clemency And Whether It Is Better To Be Loved Than Feared," gives the following advice to his Prince (published in 1532), but it should be considered in light of twenty-first-century policing.

> He ought to be slow to believe and to act, nor should he himself show fear, but proceed in a temperate manner with prudence and humanity, so that too much confidence may not make him incautious and too much distrust render him intolerable. … Therefore a prince, so long as he keeps his subjects united and loyal, ought not to mind the reproach of cruelty; because with a few examples he will be more merciful than those who, through too much mercy, allow disorders to arise, from which follow murders or robberies; for these are wont to injure the whole people, whilst those executions which originate with a prince offend the individual only.
>
> Upon this a question arises: *whether it be better to be loved than feared or feared than loved*? It may be answered that one should wish to be both, but, because it is difficult to unite them in one person, is much safer to be feared than loved, when, of the two, either must be dispensed with. Because this is to be asserted in general of men, that they are ungrateful, fickle, false, cowardly, covetous, and as long as you succeed they are yours entirely; they will offer you their blood, property, life and children, as is said above, when the need is far distant; but when it approaches they turn against you. And that prince who, relying entirely on their promises, has neglected other precautions, is ruined; because friendships that are obtained by payments, and not by greatness or nobility of mind, may indeed be earned, but they are not secured, and in time of need cannot be relied upon; and men have less scruple in offending one who is beloved than one who is feared, for love is preserved by the link of obligation which, owing to the baseness of men, is broken at every opportunity for their advantage; but fear preserves you by a dread of punishment which never fails.
>
> Nevertheless, a prince ought to inspire fear in such a way that, if he does not win love, he avoids hatred; because he can endure very well being feared whilst he is not hated. … But when it is necessary for him to proceed against the life of someone, he must do it on proper justification and for manifest cause, but above all things he must keep his hands off the property of others, because men more quickly forget the death of their father than the loss of their patrimony. Besides, pretexts for taking away the property are never wanting; for he who has once begun to live by robbery will always find pretexts for seizing what belongs to others; but reasons for taking life, on the contrary, are more difficult to find and sooner lapse.[9]

The question of love or fear is as relevant today as it was in the fifteenth and sixteenth centuries, and thus is the conundrum of police: should they be feared or loved? The quick answer is they should be respected, not from fear but from character, integrity, and performance. That is the picture we ideally want to see in reality. Instead, time after time, an officer demands respect by his verbal orders and physical appearance, thereby instilling fear through intimidation. He has the full force of the government behind him, he is empowered to take away freedom and use deadly violence, he is equipped with a gun, chemical spray, a baton, a taser, and handcuffs. He is dressed in a semi-military uniform and is wearing a bulletproof vest. "Politicians are masters at pressing our fear buttons; (sociologist Barry Glassner) quotes (President) Richard Nixon remarking, 'People react to fear, not love. They don't teach that in Sunday school, but it's true.'"[10]

To be feared or respected is the real question of policing. Fear has a negative connotation while respect has a positive connotation. Most people in power may enjoy being feared because they might confuse it for respect. There is a thin line between fear and respect with many variables like time, place, circumstances, and of course the individuals being affected. Many times fear is related to egotistical people (officers have a high egotistical quotient) who enjoy having power and manipulation over others.

The military image of police developed from the very earliest days of policing when Sir Robert Peel, the British Home Secretary, established the London Police Department. He established that police would wear uniforms with military style ranks, follow a hierarchy of a chain of command, be authorized to carry weapons and make arrests, and to use force if necessary, including deadly force. Until females began joining police departments in large numbers as patrol officers, there were strict height and weight requirements. In today's police environment, those physical restrictions have been mostly eliminated. However, the military image remains intact and with officers "bulking up" and wearing bulletproof vests, they present an impressive but often frightening and intimidating image.

A person engaged by the police and confronted by that image can be easily intimidated and fearful, often leading to a defensive stance. The demeanor of the officer can alleviate some of the fear by treating all contacts with dignity and respect. That is not to say the officer needs to say "please" (of course that is a very effective way to obtain cooperation as courtesy is more than good manners, it is a way to show respect) but attempts must be made to de-escalate the situation by tone of voice prior to resorting to "reasonable and necessary force." The degree of force escalates according to the emotions of both the police and the person being confronted.

Intimidation by a police officer can be a useful tool if properly focused. Intimidation by cursing, loud, forceful commands and posturing is not only wrong but induces fear and defensiveness in the individual being interviewed. Police tactics of threats and lies may work in some rare circumstances but in a court the case may hinge on a determination of coercion (fear) and false promises. An officer should be assertive but not aggressive and rude. A professional courteous approach will be most effective and will raise no doubts in court.

There is an old saying, "You catch more flies with honey, than you do with vinegar." This type of working with a suspect, witness, and victim will elicit more information even though the very appearance of the officer is intimidating. In the end, the office's attitude dictates responses and actions of the person with whom he or she is talking.

> Consider the emotional conditions. When an officer stops a citizen, he is entirely in control of the situation. When he handcuffs a suspect, or hurls him into a wall, he is perhaps only a little less so. When he draws his gun and points it at the ground, he

> recognizes that a danger is present but not acute; he is focused on alleviating the possibility of violence. When he raises his weapon and points it, he believes that the possibility of violence is acute but not imminent. As the situation escalates, his own fear and stress are rising, his own sense of control slipping away. When he decides to shoot, he has lost control of the situation. His fear and stress are at their maximum. It is at this point when we expect him to act the most, because the pressure on his own control is so intense, because he is acting without thinking.[11]

A humorous thought by Mark Twain gives an unexpected perspective on the topic of respect …

> Be respectful to your superiors, if you have any, also to strangers, and sometimes to others. If a person offends you, and you are in doubt as to whether it was intentional or not, do not resort to extreme measures; simply watch your chance and hit him with a brick. That will be sufficient. If you shall find that he had not intended any offense, come out frankly and confess yourself in the wrong when you struck him; acknowledge it like a man and say you didn't mean to. Yes, always avoid violence; in this age of charity and kindliness, the time has gone by for such things. Leave dynamite to the low and unrefined.[12]

The Escalation of Violence and Force Paradigm

> Every day in America, a police officer draws a gun and takes aim. What comes next is a decision of extraordinary power and responsibility. The wrong choice could end a life. Even the right decision could cause the officer a lifetime of doubt and regret. The implications are moral, political, and often racial.[13]

When a firearm is drawn the situation rises to a whole new level of danger. There are three rules of gun safety—All guns are always loaded; Never point a gun at someone unless the intention is to shoot to stop the threat; and All guns are always loaded. An officer is authorized to use his firearm if his life or the life of someone else is in immediate jeopardy. As quickly as a heart can beat once, the officer must decide if pulling the trigger is his last resort and there is no other option. Most police officers are trained in knowing the five steps of the escalation of violence and the use of force.

- **Discussion**—Simple dialogue at normal conversational tones. There is nothing to be concerned about, no flash words or red flags of dissent. If the scenario were to have a color attached to it, it would be a deep cool blue. The initial contact is very important. The voice and demeanor of the officer, conveys, or should convey, to his contact the officer's intent. A calm, friendly approach is the prescribed method for approaching an errant motorist. The traffic violator might respond in a like manner or he may become defensive and argumentative. Leading to the next stage, which is argument.
- **Argument**—Voices begin to rise, core body temperature increases, heart and pulse begin to race (adrenalin flows, testosterone joins in). Logic and reason no longer exist because neither side is listening to the other, and the emotion and belief is that the winner is the one who shouts the loudest. The cool color of blue begins to turn to yellow, orange, then red (this is where the violence begins). Both parties escalate their efforts to convince the other. Voices, body temperature, heart rate and pulse levels rise

with an infusion of adrenalin and a loss of cognitive reasoning and logic. The wisdom being that the person who shouts the loudest wins the argument and that an unreasonable person cannot be reasoned with. The unreasonable person, of course, is always the other person! However, the officer must be considered, by definition, as the person in the right. At this stage, the violator is considered as "non-compliant" or in police terminology, "hostile, belligerent, and aggressive." *There is no need to attack, if you have control of the situation*—Martial Artist and Self-Defense Instructor, Alexandr Popeskou.

- **Physical fight**—Offense is taken and no progress is made in the argument stage except the raising of the levels of adrenalin and testosterone, leading to two possible conflicting emotions: fight/flight and challenge to "manhood." Usually the fight emotion overcomes the flight (adrenalin) aspect so both sides feel a need to prove their "manhood" (the testosterone affect). The color red becomes darker, perhaps to a dark blood red, bordering on and turning black. This is the point where the officer is forced (and authorized by law) to take all reasonable and necessary action to enforce compliance.
- **Use of non-lethal weapons**—The longer a fight lasts, the greater the chance of being hurt, injured, and subdued. Tasers, batons, dirt in the face, blunt instruments, anything the combatants can grab or clutch that is not fastened down can be used to injure the opponent without killing to end the fight. At this stage "blind rage adrenalin" has reduced vision to pinpoints, leaving only one thought, "Hurt this M…er…F…er." Not being able to overcome the subject, this is the time an officer will resort to using a taser, baton, pepper spray, or other instrument short of lethal weapons. *When you disarm someone you have DOUBLE the Responsibility (your life and your opponent's life).*[14] What must be maintained in mind is that anything, such as a baton or flashlight, can become a deadly weapon, depending on how it is used. In a moment of overcoming his fear and the adrenalin dump, an officer may lose emotional control and strike too many times and too strongly, resulting in complaints of police brutality.
- **Use of lethal weapons and/or force**—A last resort to save one's own self or the life of another. This becomes the "I was in fear for my life" stage using whatever means available, usually a gun. Anyone using this defense (or offense) must be able to articulate the totality of circumstances leading to the point of being "in fear for my life." Articulating circumstances should include the progression of the escalation of force to prove it was "reasonable and necessary." Explanation should also include the intent of the shooting was to "stop the person according to department policy and training."

It must be considered that anything can be a lethal weapon, depending on how it is used. Most commonly the reference is to the use of an officer's sidearm. Even then there are strict legal guidelines that must be followed: (1) It is to be used only as the very last resort; (2) the weapon must be used only to protect the officer's life or the life of another in immediate danger; (3) The officer must be able to articulate the totality of circumstances of why and how he feared his life or the life of another was in immediate danger.

I Was in Fear for My Life …

"I was in fear for my life" has become as common a statement as any cliché like, "He went to his waistband" or, "He made a sudden move with his hand." The first rule of gun safety, "All guns are always loaded," must be uppermost in an officer's mind because to make an assumption or to wait can be deadly. However, there is a difference in *thinking*—through the blurred vision of the effects of adrenalin—that an individual might have a weapon from actually *knowing*. The officer's mindset must be that any circumstance can turn violent and deadly in less time than it takes to inhale a breath. Extreme caution and awareness are principles of protection for the safety of officers. On the other hand, if no weapon is found and the officer cannot explain what led him to believe his life was in immediate danger, then the excuse of "I was in fear for my life" should not be a viable reason for deadly force.

Fear Is a Liar!

Fear is a liar. Fear distorts the truth of reality with imperfect perception. There have been many instances of police shootings when the officer said, "I was in fear for my life" when the circumstances clearly do not sustain the claim. One instance is very vividly illustrated in a police shooting in the back of a fleeing man on April 4, 2015 in North Charleston, South Carolina. A video of the incident clearly shows the fleeing man several feet in front of the officer when the officer shoots at him eight times, with at least three of the bullets hitting and killing the man. The incident began, according to the police officer, when he stopped the man for a minor traffic violation (a brake light out). A warrant check revealed the man had warrants outstanding for delinquent child support payments. The man ran away and the officer opened fire. The officer claimed the man had stolen his taser. Three days later, the officer was arrested and charged with murder[15] (see Chapter 12).

The following was posted by *Common Dreams* on June 25, 2015:[16]

> Police kill a lot of unarmed people. So far in 2015, as many as 100 unarmed people have been killed by police. Here are fifteen of the most outrageous reasons given by police to justify killing unarmed people in the last twelve months.
>
> First, a bit of background. So far in 2015, there have been around 400 fatal police shootings already; one in six of those killings, 16 percent, were of unarmed people, 49 had no weapon at all and 13 had toys, according to the Washington Post. Of the police killings this year less than 1 percent have resulted in the officer being charged with a crime. The Guardian did a study which included killings by Tasers and found 102 people killed by police so far in 2015 were unarmed and that unarmed Black people are twice as likely to be killed by police as whites.
>
> **One.** He was Dancing in the Street and Walking with a Purpose. On June 9, 2015 an unarmed man, Ryan Bollinger, was shot by police in Des Moines after "walking with a purpose" towards the police car after he exited his vehicle after a low speed chase started when he was observed dancing in the street and behaving erratically. The deceased was shot by the police through the rolled up cruiser window. The murder is under investigation.
>
> **Two.** Thought It Was My Taser. An unarmed man, Eric Harris, ran from the police in Tulsa Oklahoma on April 2, 2015. After he was shot in the back by a Taser by one officer and was on the ground, another 73 year old volunteer reserve officer shot and killed him, all captured by video. While dying he was yelling that he was losing his breath, to which one of the officers responded "F*ck your breath." The police said the officer thought he was shooting his Taser and "inadvertently discharged his service weapon." The officer has been charged with second degree manslaughter. Running away from the police so often provokes police overreaction that the aggressive police response has several names including the "foot tax" and the "running tax."
>
> **Three.** Naked Man Refused to Stop. A naked unarmed mentally ill Air Force Afghanistan veteran, Anthony Hill, was shot and killed March 9, 2015 by DeKalb County Georgia police after police said he refused an order to stop. The killing is under investigation.

Four. Not Going to Say. On March 6, 2015 Aurora Colorado police shot and killed unarmed Naeschylus Vinzant while taking him into custody. For the last three months, while the investigation into the killing continues, the police have refused to say what compelled the officer to shoot Vinzant.

Five. Police Felt Threatened by One Unarmed Homeless Man. March 1, 2015 Los Angeles police shot and killed an unarmed homeless man Charly Leundeu Keunang after five officers went to his tent and struggled with him. One unarmed homeless man threatened five armed LAPD officers? Los Angeles police have killed about one person a week since 2000. An investigation is ongoing.

Six. My Taser Didn't Work. On February 23, 2015, an unarmed man, Daniel Elrod, was shot twice in the back and once in the shoulder and killed by Omaha Nebraska police after he tried to climb a tree and jump a fence to escape police who suspected him of robbery. Police said their Taser did not work, he ignored their demands to get down on the ground, he did not show his hands, and they felt threatened. Video was not made available and the officer later resigned. This was the second person this officer killed. No criminal charges were filed.

Seven. Armed with a Broom. Lavall Hall's mother called the police in Miami Gardens February 15, 2015 and asked for help for her son who was mentally ill. Lavall Hall, five foot four inches tall, walked outside with a broom and was later shot and killed by police who said he failed to comply with instructions and engaged them with an object. The killing is still under investigation.

Eight. Throwing Rocks. On February 10, 2015 an unarmed man, Antonio Zambrano-Montes, was fired at 17 times and killed by police in Kennewick, Washington. A video of his killing has been viewed more than 2 million times. Officers said he had been throwing rocks at cars, ran away and then turned around.

Nine. Taser Worked but He Didn't Stop Moving. On February 2, 2015, a Hummelstown Pennsylvania police officer shot unarmed David Kassick in the back with a Taser and when Kassick went to the ground on his stomach, then shot him twice with her gun in the back, killing him. The officer said Kassick, who was running away from a traffic stop, was told to show his hands and not move but continued to try to remove the Taser prongs from his back and the officer thought he was reaching for a gun. The officer has been charged with homicide.

Ten. Car going 11 Miles an Hour was going to Kill Me. Denver police fired 8 times at unarmed Jessica Hernandez, 17, who was killed January 16 after being hit by four bullets. The police said she drove too close to them when she was trying to get away and may have tried to run them down as she tried to drive away so they shot into the windshield and driver's windows. The police said the car may have reached 11 miles per hour in the 16 feet it traveled before hitting a fence. The police were not charged.

Eleven. Armed with a Spoon. Dennis Grigsby, an unarmed mentally ill man holding a soup spoon, was shot in the chest and killed in a neighbor's garage by Texarkana Police December 15, 2015. The killing is under investigation.

Twelve. Armed with Prescription Bottle. Rumain Brisbon, a 34 year old unarmed man, was shot twice and killed by police in Phoenix on December 2, 2014, after he ran away, was caught and was in a struggle with the officer who mistook a prescription pill bottle in Brisbon's pocket for a gun. The police officer was not charged.

Thirteen. It Was an Accident. On November 20, 2014, a New York City police officer fired into a stairwell and killed unarmed Akai Gurley. The officer, who was charged with manslaughter, is expected to say he accidently fired his gun.

Fourteen. Don't Mention It. On November 12, 2014, an unarmed handcuffed inmate was shot multiple times in the head, neck, chest and arms by officers while fighting with another handcuffed inmate in the High Desert State Prison in Carson City Nevada. His family was not told and did not know he had been shot until three days later when they claimed his body at a mortuary.

Fifteen. Armed with Toy Gun. John Crawford was unarmed in a Walmart store in Beavercreek Ohio on August 4, 2014, when he picked up an unloaded BB gun. When officers arrived they say they ordered him to put down the gun and started shooting, hitting him at least twice and killing Mr. Crawford. In a widely viewed video Mr. Crawford can be seeing dropping the BB gun, running away and being shot while unarmed. Likewise, Cleveland police shot and killed an unarmed 12 year old boy, Tamir Rice, who was playing with a toy pellet gun on November 22, 2014. Police said they shouted verbal commands from inside their vehicle in the two seconds before they shot him twice. In both these cases, the police story of shouting warnings and orders looks quite iffy at best.

These are the responses of police authorities who face less than one chance in a hundred of being charged when they kill people, even unarmed people. These outrages demand massive change in the way lethal force is used, reported, justified and prosecuted.

All of the above small samplings of police shootings are either by tragic mistakes or the officers claimed to have been in "fear for my life." With accounts of these shootings reported by mass media of news outlets and social Internet sources, the public has become concerned by the often unwarranted and possibly illegal shootings of citizens by police who make a claim of "I was afraid for my life." If officers so easily become afraid, they may be in the wrong business. Every police encounter with citizens holds the promise of potential harm.

Of course these shootings may be anomalies but they have similarities leading noninvolved people to conclude the officers responded with lethal force because of fear.

Attitude

The driver looks into his rear-view mirror and sees red and blue lights of a police patrol car behind him, telling him to "pull over and stop." At this point the driver begins to experience an emotional and mental transformation. Whatever he was thinking about before the police lights came on is suddenly disrupted by sudden chemical changes in his body. Adrenalin pushes through his system and fearfulness becomes part of his thought process as he mentally curses, "Damn it, What did I do?" or "This is going to cost me money," or "Oh, oh, a police officer, what do I say?" Emotions run from fright to defensive.

A person walking down the street is stopped by a passing police car. The officers approach with caution and require identification and a stated purpose of why the person is walking there, where are they going, and where are they coming from. The person who is confronted is confused, perhaps frightened because he was simply and innocently walking home from the store. He is unaware of the U.S. Supreme Court decision of *Terry V. Ohio*, which is that if an officer has "reasonable suspicion" that a crime has or might be committed, based on his experience and facts of his observations that he can fully articulate, he may stop a person and "pat" him down or "frisk" him for weapons and make the appropriate inquiries. Emotions of the person being stopped and questioned run from fear to defensive.

Partygoers in a private residence in a usually quiet neighborhood are making so much noise that the neighbors call the police to "stop that racket from next door." As the police car parks at the curb and the officers begin to walk toward the house, they are seen and a partygoer shouts, "Here come the cops." People rush to hide whatever they are smoking or ingesting and put on a "sober" face. Attitudes and emotions run from fear to defensive.

Quoted here in its entirety is one person's negatively jaundiced perspective, viewpoint, and attitude toward the police. Obviously this person has very prejudiced and cynical reactions to government in general, and police in particular. What is surprising in this interview is that the person is a highly talented film producer and creative artist who normally projects an image of being a law-abiding citizen. (Please note, this interview was conducted with a personal friend of the author, who wishes to remain anonymous.)

> Yes, I've had a run-in with the police. Surprise, surprise—it didn't happen until I moved to North Carolina. The "probable cause" went from my tires "squealed" to I "skidded out" to when I went to court, I "spun out my vehicle," which is horseshit. Yes, I happened to have marijuana, which is a very very rare occurrence. However, the officer's falsification of events reaffirmed to me what I already know.
>
> To me, police officers are the corrupted enforcers of the elite—those in power who themselves are the ultimate hypocrites, and who enslave the underclass of the less privileged. Anybody with open eyes can see that the government "leaders" are in the pockets of the corporations—those who actually run this nation, and they're complicit in taking kick-backs for securing or undoing legislation which furthers the profiteering of said corporation, which line their pockets.
>
> In my eyes, the "cops" are the same individuals who were the bullies in elementary school—rejected, dejected and attention starved. Usually they are the lower IQ subordinates, who feel it is justifiable to continue "just following orders," without the mental capacity for objective or rational thought—which is the reason they're hired for the job in the first place.
>
> I see a cop as something to be avoided with prejudice, like a deadly rockfish or a festering Ebola-like disease. I have no tolerance nor compassion for them, instead viewing them as a robot-like obstruction, and yes, I do wish harm upon them. They are the first line in the oppression, destruction and sorrow of mankind, and I detest them like no other being (save for the politicians who pull their puppet strings).
>
> Individuals without the capacity for objective thought are to me nothing more than enemy combatants, deserving of a horrible, brutal and excruciatingly prolonged death. I have nary a higher level of contempt than I do for law enforcement. Interesting, though, I was stopped in Montana the other day for driving 102 mph and the cop wasn't a dick at all. This happened once before in 2003, and same deal. I actually asked the officer why, and he said "less stress" there.
>
> The world would be a better place without roving packs of violent armed mongoloids roaming the streets extorting money and ruining lives. They're nothing different than criminal gangs, government sanctioned, and I wish for them a hell ten times beyond the one they're inflicting daily upon our citizens. And here I'm not padding the truth but speaking lightly.

In all of these scenarios, and every other one, from a simple traffic stop, even to giving a warning, the officers know violence can reach out any moment and they must be mentally alert for any indications of anything that may possibly happen to turn to the use of force. At the same time officers must recognize that their approach and demeanor normally set the scale of the attitude of the person they have contacted.

In a July 2015 incident in Hempstead, Texas, a young woman (Sandra Bland) was stopped by an officer of the Texas Department of Public Safety for failing to signal when changing lanes on the highway. Three days later she was found dead in her jail cell, hanging with a plastic trash bag around her neck. The official autopsy report listed the cause of death as asphyxiation by suicide.

While questions about her death were being asked, the Waller County District Attorney was quoted as describing (the woman) as "Very combative and non compliant," but he also went on to say it also appeared the officer was also at fault, noting, "It was not a model traffic stop and it was not a model person that was stopped." The arresting officer was placed on desk duty after a preliminary review showed violations of traffic stop procedures and "courtesy policy," according to a statement from the Department of Public Safety.[17]

A review of the officer's vehicle dash camera gives a valid impression that the entire incident turned into a physical confrontation when the officer presented the woman with a citation he later claimed was a "warning." He then asked her if she was upset. She replied in the affirmative and began to angrily tell him why. The officer told her to put out her cigarette and she refused. He then ordered her out of the car and, when she refused, he opened the door and attempted to pull her from it, and told her she was under arrest. The question is, did the officer act the way he did because of ego or an enforcement issue? There was no legal issue, he had already presented the warning citation which was his reason for stopping her. The incident escalated when he ordered her to put out her cigarette and to get out of the car. There was also no probable cause or reasonable suspicion to have her exit the car and she had the right to refuse. There was no legal grounds for an arrest.[18]

Attitude is a powerful force. In the incident above the matter turned to violence when the officer apparently felt his authority and police power allowed him to instruct the woman to put out her cigarette and get out of the car. Even prior to that, when the woman was explaining she was angry and upset about getting a citation for not signaling a lane change on a nearly empty highway (she had not been told it was only a warning), the officer should not have taken it as a personal affront to him. Being verbally disrespectful to an officer is not a crime. When he handed her the citation, the police work was finished and he should have simply turned, let her go on her way, and walked back to his car—contact finished—case closed.

Reflective in this case is a majority decision of the Supreme Court in the case of *Rodriguez V. United States* decided April 21, 2015. Justice Ruth Bader Ginsburg wrote that

> [T]he tolerable duration of police inquiries in the traffic-stop context is determined by the seizure's "mission"—to address the traffic violation that warranted the stop, and attend to related safety concerns. [A police stop] may "last no longer than is necessary to effectuate th[at] purpose." Authority for the seizure thus ends when tasks tied to the traffic infraction are—or reasonably should have been—completed. [...] If an officer can complete traffic-based inquiries expeditiously, then that is the amount of time reasonably required to complete the stop's mission and he may hold the individual no longer.[19]

But what is clear now, from the Sandra Bland traffic stop video, is that there was no other crime, nor a safety concern. The officer was acting in violation of the law, as defined by the Supreme Court. It is incidents like this that cause fear of the police by the public.

I have been associated with law enforcement my entire adult life and have seen way too much of negative types of officers' attitudes and egos. If police are "our friends" they must earn our respect worthy of that trust. "Treat everyone with dignity and respect—Courtesy is a way to show respect to others." The large majority of police understand this and follow that line, but too many allow their egos and power mania to interfere and they lose all control of themselves, making the scenario more dangerous.

The Attitude Test

> More people talk their way into jail (or a citation) than out of it.
>
> *Unwritten police philosophy*

In police parlance there is an unwritten, unspoken procedure called "the attitude test." There is no secret formula or criteria for the test. It begins when a police officer makes initial contact with an individual. In an ideal situation if the officer feels the person is cooperative and non-aggressive, his or her own personal demeanor should be reflective—professional but friendly. On first beginning the contact, officers should greet the individual with courtesy and a simple "Hello, Sir, I stopped you because …" This should set the tone of the interview in a non-confrontational way and lead to an exchange of information under no duress. If the individual becomes defensive, uncooperative and argumentative, the officer, who may have not planned on citing the person or was planning on issuing a warning, will become aggressive and assertive with a growing escalation through the five stages of the use of force, leading to arrest violence and death. The statement "more people talk their way into jail than talk themselves out" becomes true.

If the officer initiates the contact in a surly, disrespectful manner, he or she can expect to be met with the same attitude, quickly causing a potentially explosive circumstance. This is not to say an officer should not be cautious when approaching a vehicle or a pedestrian. Nor should he or she be non-assertive. A major key is to be friendly and respectful while maintaining an official position.

Officers must feel confident they are trained correctly to act when appropriate, that the training they get will save their lives, and that their department/community will support them.

> Policing is damn hard, but we have to put an end to the killing of unarmed citizens who are not putting officers at risk. I always try to separate fear from danger, because it is danger that puts officers at risk, but it is often fear that causes the public to be severely injured or killed. Every time a person experiences fear, it does not mean they are in danger.[20]

Summary

Fear is a liar! It distorts reality. Anger, the lion of emotions, overrides other emotions such as depression and fear. Anger arousal transforms a docile individual into a stark raving and fierce fighting machine. An angry person shows no regard for consequences either to himself or others. He becomes steeped in aggression, feeling a hot burning rage, the color of red being the only thing in his vision. In times like this he will allow all reasoning and common good judgment to become overridden with only one thought, destroy the person or thing causing the fear.

The many experiences of a police officer witnessing the worst of society, often called "sleaze," injustice, heartbreak, and violent death, builds to high pressure, stress, and PTSD. As frustration grows, taking its toll, and emotional control loses its power, the officer may react with anger over what may be considered by some as a minor incident. He may react viciously and out of control with rage, resulting in savagely beating or killing the source of his fear and anger. He may then attempt to justify or rationalize his behavioral anger and fear by believing "I was in fear for my life." In his novel *Oliver Twist*, Charles Dickens underscored the thinking of some police officers: "To do a great right, you may do a little wrong; and you may take any means which the end to be attained will justify."

There are five steps in the model for escalating use of force, beginning with simple discussion or the mere presence of an officer and ending with the last resort of using lethal weapons.

The two most likely threatening situations in the "routine day" of a police officer are responding to a domestic violence dispatch or confronting a person or persons who refuse to comply with his orders or instruction. Over a period of time the officer may begin to experience "anxiety dreams," presenting him with anxiety that he may engage in a physical fight or gunfire in the domestic violence and non-compliance and not be able to react properly.

Review Questions

1. In your own words explain the acceleration of aggression.
2. Is it better for an officer to be feared or loved?
3. Discuss ways police are intimidating.
4. What is the difference between assertive and aggressive?
5. How is Fear a liar?

Notes

1 Hayes, Laura L.; "How to Stop Violence: Mentally Ill People Aren't Killers. Angry People Are"; April 9, 2014; www.slate.com/articles/health_and_science/medical_examiner/2014/04/anger_causes_violence_treat_it_rather_than_mental_illness_to_stop_mass_murder.html. Retrieved September 26, 2018.
2 King James Bible; John 15:13.
3 www.brainyquote.com/quotes/david_horsey_714234?src=t_anger. Retrieved February 12, 2019.
4 www.listofhumanemotions.com/anger. Retrieved July 16, 2015.
5 www.webmd.com/mental-health/tc/fear-and-anger-in-ptsd-overview. Retrieved April 16, 2017.
6 Shakur, Assata; *Assata: An Autobiography*. Originally published in the UK by Zed Books (1987). Published in the United States by Lawrence Hill Books (foreword by Angela Davis).
7 Assata Olugbala Shakur, 67, is an African-American activist and member of the former Black Panther Party and Black Liberation Army. Between 1971 and 1973, Shakur was accused of several crimes and was the subject of a multistate manhunt. She became a member of the Black Power movement at a time when many activists were galvanized following the assassination of Dr. Martin Luther King. In 1973, Shakur was arrested during a traffic stop on the New Jersey Turnpike. A shootout left Assata injured with multiple wounds, the driver, Zayd Shakur, dead, and a state trooper dead. In 1977, after numerous trials, she was convicted of first-degree murder of the officer. And in 1979, Shakur escaped and fled to Cuba, where Fidel Castro granted her asylum. Love, David, A. "Will Cuba Extradite Assata Shakur to the US?": December 18, 2014 https://thegrio.com/2014/12/18/assata-shakur-us-cuba. Retrieved July 25, 2015.
8 Shakur, Assata; *Assata: An Autobiography*, Originally published in the UK by Zed Books (1987). Published in the United States by Lawrence Hill Books (foreword by Angela Davis).
9 Machiavelli, Niccolò. *The Prince;* Chapter XVII; "Concerning Cruelty and Clemency, and Whether It Is Better To Be Loved than Feared"; www.constitution.org/mac/prince17.htm; Retrieved July 17, 2015.
10 www.kirkusreviews.com/book-reviews/barry-glassner/the-culture-of-fear/. Retrieved December 8, 2018.

11 Wallace-Wells, Benjamin; "Police Shootings, Race, and The Fear Defense"; *The New Yorker*, July 12, 2016.
12 www.thoughtco.com/advice-to-youth-by-mark-twain-1690213. Retrieved May 11, 2018.
13 Sloan, Steven; "The Trigger and the Choice"; at www.cnn.com/interactive/2017/politics/state/kyle-dinkheller-police-video/. Retrieved January 2, 2019.
14 For further information see: Alexandr Popeskou #berkootlife #systema #personalprotection. Retrieved November 4, 2018.
15 www.cnn.com/2015/04/08/us/south-carolina-cop-shoots-black-man-timeline/index.html. Retrieved July 19, 2015.
16 Quigley, Bill; Fifteen Most Outrageous Responses by Police after Killing Unarmed People; published on Thursday, June 25, 2015; www.commondreams.org/views/2015/06/25/fifteen-most-outrageous-responses-police-after-killing-unarmed-people. Retrieved October 17, 2019.
17 Hennessy-Fiske, Molly; *Los Angeles Times*; "Answers Sought in Jail Cell Death." July 21, 2015, p. A6.
18 For further information regarding this incident: www.bing.com/search?q=sandra+bland+waller+county&qs=PS&sk=AS1&pq=sandra+bl&sc=8-9&sp=2&cvid=a5231f89885a497c802b41e315181b3b&FORM=QBLH&ghc=1.
19 www.supremecourt.gov/opinions/14pdf/13-9972_p8k0.pdf. Retrieved July 25, 2015.
20 Mark James—President of Panther Protective Services—www.pantherprotectionservices.com/about.

14

POLICE SUBCULTURE AND PERSONALITY

Cultures

Police culture is peppered with stereotypes. There is a jokey stereotype of police culture. "They will pass everything on the street except a donut store."

> Frosting was his favorite. He liked to eat doughnuts at every meal. Because it was healthier to eat six small meals a day than three large ones, he restricted himself: jellied for breakfast, glazed for brunch, cream-filled for lunch, frosting for linner, chocolate for dinner, and powdered sugar for 2 a.m. supermarket stakeout. Because dinner coincided with the daily crime peak, he always ate his favorite variety to ease him. Frosting was his only choice now, and upsetting his routine was a quiet thrill.
>
> *Benson Bruno*[1]

Culture is defined by a standard dictionary as: the characteristics of a particular group of people, defined by everything from language, religion, cuisine, social habits, music, and arts; The set of shared attitudes, values, goals, and practices that characterizes an institution or organization.

A subculture is an ethnic, regional, economic, or social group exhibiting characteristic patterns of behavior sufficient to distinguish it from others within an embracing culture or society having a unique language (jargon and slang), values, rules, and judgments, for example: *a criminal subculture or a police subculture. A subculture is a deviation from the accepted norm. Deviation is not a crime but a crime is deviation.*

> The informal values of police subculture have been linked with a variety of forms of police deviance, including the abuse of discretion and authority, use of dishonesty as a tool, accepting bribes and payoffs, brutality, and the use of excessive (or unnecessary) force and perjury. As is the case in other professions, socialization into the informal values of an organization can force a professional to abandon previously held ethical standards or, at the very least, make it difficult to uphold personal moral ideals as well as the formal values of the profession."[2]

In most sociology books, a culture is defined as "The beliefs, customs, arts, etc., of a particular society, group, place, or time; a particular society that has its own beliefs, ways of life, art, etc.; a way of thinking, behaving, or working that exists in a place or organization." A subculture is commonly defined as "An ethnic, regional, economic, or social group exhibiting characteristic patterns of behavior sufficient to distinguish it from others within an embracing culture or society, *i.e. police subculture.*"

Subculture is a form of institutionalization. It is:

> the process of embedding some conception (for example a belief, norm, social role, particular value or mode of behavior) within an organization, social system, or society as a whole. It is the process which translates an organization's code of conduct, mission, policies, vision, and strategic plans into action guidelines applicable to the daily activities of its officers and other employees. It aims at integrating fundamental values and objectives into the organization's culture and structure.[3]

In every society there are innumerable subcultures: motorcycle gangs, street gangs, fraternity and sorority groups, fire fighters, military, etc., and police.

> The strength of the [police] subculture grows out of the peculiar characteristics and conflicting pressures of the job; the ever-present physical danger; the hostility directed at the police because of their controlling role; the vulnerability of police officers to allegations of wrongdoing; unreasonable demands and conflicting expectations; uncertainty as to the function and authority of officers; a prevalent feeling that the public does not really understand what the police have to "put up with" in dealing with citizens; a stifling work environment; the dependence that officers place on each other to get the job done and to provide for their personal safety; and the shared sense of awareness, within a police department, that it is not always possible to act in ways in which the public would expect one to act.[4]

Subcultures have their own language, commonly referred to as "jargon." A good part of transitioning from a private citizen to an officer is learning the specialized vocabulary and integrating into the subculture.

Police jargon includes: acronyms, abbreviations, codes, and slang used by law enforcement personnel to provide quick, concise descriptions of people, places, property, and situations. Some simple expressions are: DOB (date of birth); DOA (dead on arrival); "Roger that" (I understand); 10–4 (acknowledge); Mope (same as sleaze bag or dirt bag—someone held in disrespect); Perp (perpetrator); "F him" (California Penal code 647f—to arrest for drunk in public and unable to care for himself or others).

Developing a Police Personality

> The characteristics usually associated with police personalities in present times are machismo, bravery, authoritarianism, cynicism and aggression. Additional characteristics have been associated with police personalities as well: suspicious, solidaristic, conservative, alienated and thoroughly bigoted.[5]

Like military basic training transforms a civilian into a soldier, a police recruit training academy has a similar effect on a recruit: discipline, physical training, police tactics, principles of law enforcement and criminal law turn an accepted candidate into a police officer. Candidates are selected according to four I's; Intelligence, Integrity, Imagination and Interpersonal skills.

- **Intelligence** is the ability to learn and understand. It includes cognitive analysis and critical thinking. Critical thinking is an analysis through the asking of questions and the employment of deductive reasoning. It is a basic tool of investigators, researchers, writers, students and professionals. When all facts and circumstances are gathered and organized, the critical thinker (through logic and deduction) is able to arrive at a conclusion commensurate with the facts. Critical thinking tools include emotional feelings, visual images, bodily sensations, reproducible patterns, observation, education, perspectives, reasoning and analogies. A good critical thinker can translate these tools into thoughts expressing insight that can give rise to new ideas in others' minds. ' Critical thinking is that mode of thinking about any subject, content, or problem—in which the thinker improves the quality of his or her thinking by skillfully analyzing, and reconstructing it. Critical thinking is self-directed, self-disciplined, self-motivated, and self-corrective thinking."[6] It is that ability to view a particular set of circumstances from a number of sources and perspectives to form an inference based on analysis, reasoning, logic, experience, and training. It is a synonym for ensuring that you account for all significant consequences by developing the habit of using your imagination. Critical thinkers are "imaginators," having the ability to look at something and see more than the physical representation.[7]
- **Integrity** is morality, being ethical, having trustworthiness, and honesty beyond question.
- **Imagination** is a valuable trait in solving questions and crimes. It is the ability to ask, "What if ...?" and to form a theory or create something new; having original thoughts and coming to plausible conclusions.
- **Interpersonal skills** are the ability to reason effectively to convince and to resolve conflicts with opposing parties. It is being more pragmatic than dogmatic (willing to compromise and to see an opposing viewpoint).

Lost in stressful, adrenalin-filled moments and given little attention in police training is an important qualification sought in police recruiting—interpersonal skills.

> According to a 2018 article by Timothy Roufa, a law enforcement subject matter expert, "In some of those encounters, tensions may be high and the potential for the situation to quickly escalate is a reality of the job. The best way to resolve potentially dangerous use-of-force situations is to rely on your cognitive and emotional intelligence skills. [...] Roufa further explains that de-escalation does not involve the hard skills acquired through professional training, but rather the soft skills you need to develop to be truly effective in your day-to-day life as a police or corrections officer ... Roufa highlights the importance of the following soft skills (interpersonal skills) for criminal justice professionals:
>
> **Empathy**—The ability to not only understand someone else's feelings but also to share those feelings, which allows for a deeper appreciation of what others are experiencing.

Compassion—Compassion begins where empathy leaves off. If empathy is an understanding and sharing of others' feelings, then compassion means putting that understanding into action.

Nonverbal Communication—Those cues we send through our tone of voice, facial expressions, gestures, and enunciation often carry far more weight in how our messages are received than the actual words we use.

Active Listening—In our profession, most situations involve individuals who just want to be heard. Active listening means interpreting and understanding the needs of others in a conversation.

Adaptability—Anyone who works in law enforcement or corrections will tell you that your day is going to be largely unpredictable; we must be able to anticipate, adapt, and overcome challenges that we are likely to encounter.

Building Rapport—As mentioned, we will encounter a number of individuals each day, each with their own wants and needs, so we must establish a rapport with these individuals in order to build a certain level of mutual trust.

Critical Thinking and Observation—There is no such thing as a routine day when you work in law enforcement. Officers must be able to quickly and efficiently assess and analyze facts, observations, and information so they can make informed, sound decisions. Critical thinking is essential to solving problems and resolving conflicts. Keen observational skills require that officers be able to visually, mentally, and emotionally gauge a situation, often quickly. Officers must be detail-oriented, so they can observe and identify small, yet important details.

Conflict Resolution—Because conflict accompanies much of an officer's job (corrections and law enforcement), he or she must have the ability to resolve that conflict quickly and peacefully.[8]

The process of socialization into the (police) culture begins as soon as new recruits enter the academy. They get to know not only the formal rules of policing but also the informal norms that a person has to abide by in order to be accepted into the group. They learn very quickly that loyalty—the obligation to support a fellow officer—is the first priority.[9]

This philosophy has become known as "the blue wall of silence." In more colloquial terms, the meaning is "if you lie, I'll swear to it."

It is obvious in modern policing that there is no room for the following so-called principles. However, it is now a recognized fact that they are still regularly practiced and commonly adhered to. The "blue wall of silence" and the "power syndrome" are inherent in the informal police values system in the subculture. Lawrence Sherman identified eight core informal rules (1982) that police socialization and institutionalization have developed counter to good ethical practices and the law enforcement code of ethics.

1. Decisions about whether to enforce the law "should be guided by both what the law says and who the suspect is. Demographic variables such as the race, class, and age of the suspect, as well as the suspect's demeanor, attitude, and level of cooperation should all be taken into consideration." (This is commonalty known as "the attitude test").
2. Disrespect for the police should always be met with arrest or the use of force. Even when no law violation has occurred, the officer should find a way to impose punishment—even if this means making an arrest on fake charges.

3. Officers should not hesitate to use force—including deadly force—against those who "deserve it" or where it might otherwise be an effective means of solving a crime. When she or he can get away with it, the officer should employ "as much force as society should use on people like that—force and punishment which bleeding heart judges are too soft to impose.
4. Due process exists for the sake of protecting criminals and should be ignored whenever the officer can get away with it. (Get the contraband and the suspect off the street regardless of the legality of the arrest.) Ignoring Miranda (Fifth Amendment) rights, illegal searches and interrogations and the use of physical force to coerce confessions are all acceptable ways to achieve the goal of fighting crime.
5. Dishonesty—lying and deceptive methods—is "an essential part of the police job." Even perjury should be used if necessary to protect oneself or to get a conviction. Lying in court, to drug dealers, prostitutes, muggers, and burglars are legitimate means of investigating crimes and catching criminals. (The term for lying in court testimony is "testilying.")
6. "You cannot go fast enough to chase a car thief or traffic violator nor slow enough to get to a 'garbage' call; and when there are no calls for service, your time is you own." Hot pursuits are necessary to catch those who challenge police authority, social work problems such as domestic disputes are not important, and when there are no calls, one can sleep, visit friends, or anything else one can get away with.
7. It is acceptable to take any extra rewards the public wants to offer an officer. Policing is dangerous work done for low pay, so free meals, gifts, regular payments for special treatment, taking money for not giving a ticket, etc. are legitimate forms of additional compensation.
8. Protecting fellow officers is the "paramount duty"—even if it means risking your own job or life. An officer should do everything he or she can do to protect fellow officers, including never cooperating in an investigation against another officer and never "blowing the whistle" by informing on an officer who has taken a bribe, used excessive force, etc.[10]

These traits become obvious as the rigor of training begins to separate the "wannabes" from the future officer. With each passing week a new confidence and swagger become evident as is a feeling of *esprit de corps* among the trainees. That enthusiastic feeling carries over into the traditional dynamics of his or her department.

One recruitment deputy from a large sheriff's department told potential recruits:

> Our training is based on a system not unlike military basic training. When you graduate from our academy you will have learned to have no fear and gained confidence that you could go to any agency in the nation and be hired because our training is recognized as one of the most rigorous in the country and our training reputation is known for turning out the very best officers.

To many, this short paragraph is a siren song speaking of excitement, adventure, super powers, and challenges.

Attitude Timeline of Police Work

Studies have determined that an officer's mental state and attitude pass through several phases during his or her career; from excitement and fascination to a cross-roads crisis of awaiting retirement and confused about what his or her life will be like upon retirement. Most studies break the time frames into four phases of 1–5 years; 6–10 years, 11–15 years, and 16 years to retirement. Many of the phases overlap and the involved years vary according to the individual officer. This is not to say that all officers exhibit the same attitude changes, but as the years go by and the officer reaches middle age he or she becomes more cautious and wary of potential threats to his or her eventual safe retirement.

The first phase, usually the first 5 to 6 years, is like an employment "honeymoon," the new officer relishes his role, and he believes his mission is to "help others." His entire life is absorbed by being involved in his police job. He is consumed with the excitement of police work, idealistic, excited, he is fearless, "gung ho," and eager—volunteering for overtime, extra duty, and special assignments; believing it is his duty, his responsibility to ensure peace and safety in the city. He purchases the latest ancillary gear and high-powered flashlights, and other gadgets he believes will turn him into a "super cop." He arrives to work early and leaves late, never concerned about overtime. He can't wait to begin his patrol duties on the street and believes he is invincible. His social friends outside of the department outnumber his law enforcement friends but slowly the number of his outside friends begins to dwindle as the focus of most of his attention and conversation is on his police work, and more and more time is spent with fellow officers at social affairs.

The second 5 years as a law enforcement officer finds the officer settling into the everyday routine of his career which he is now finding boring and everything becomes repetitious. His eagerness to work turns to dread and he arrives to work within minutes of his duty time; idealism slowly turns to cynicism, and he adapts a suspicious attitude about everyone who is not a police officer. He has established a daily routine; he goes to work every day with no special expectations except to make it through the day with no unusual incidents that will make him work overtime or do excessive paperwork. The excitement has passed, boredom of the routine sets in, and he doesn't volunteer for new assignments or promotion. He no longer volunteers for overtime because he works at a second "off duty job," which pays more than the overtime. He no longer takes unnecessary risks and is no longer fearless nor invincible. His approach to a situation is cautious and defensive but feels he has to prove he is still able to physically subdue any offender, sometimes reacting with excessive and unnecessary force showing he is unafraid.

From 10 to 15 years in law enforcement the officer is dealing with disillusionment. He becomes cynical, distrusting, complains a lot, feels unappreciated; stress leads to sickness (particularly stomach, heart, and liver problems). He is critical of supervisors and develops an "us versus them" attitude about his supervisors and the public. He complains about his supervisors and co-workers, perhaps his marriage and family life is edging toward dissolution, he might begin using alcohol on a regular basis, visiting and spending time in a bar after work even when getting off shift in the early mornings. He may often have a drink or two while on the job. He becomes depressed and reacts with anger, quickly taking it out on the "douche bags, punks, mopes, scumbags, hookers, druggies and freeloaders." His disillusionment and cynicism make him feel unappreciated. Racism and sexism become rampant, becoming part of his personality. He has an unhappy family and social life with his only social network consisting entirely of other officers in the same career phase. If he has any friends, they are all select fellow police officers. His assertive approach has developed into aggression to prove he is still relevant, and he may counter any fear with a volatile temper, over-reacting with violence.

From approximately 16 years to final retirement the officer is at a crossroads, not wanting anything that will interfere with his plans to retire or endanger his pension. He begins experiencing the so-called "midlife crisis," knowing he has reached the apex of his career, expecting no more promotions. He merely shows up to work to put in his time, doesn't volunteer for any assignments, and senses his mortality. His focus is centered on what occupation he will engage in after retirement. Or, if he is still on patrol, if he will reach retirement age without being injured or killed, and he hopes to spend his working hours assigned to a desk job. In the meantime, he continues working, simply putting in time, collecting his salary, and waiting until that day when he will substitute his regular pay for his retirement pay.[11]

The Blue Code of Silence

Historically, law enforcement officers have considered themselves as a fraternity, a brotherhood and family. An unwritten bond grows among them, demanding one member does not become a "whistleblower" against another when observing lying, theft, excessive force, corruption, or other misconduct. The code is a very strong pillar of police subculture and an influential factor in the police personality. Integrity is an important element in recruitment and training of new officers. Unfortunately, the code is ingrained in traditional policing and undermines most efforts to eradicate it through ethics training. Slowly, like peeling the layers off an onion, a new officer is introduced to "You don't 'rat,' 'cop,' 'fink,' or tell on" your fellow brothers or sisters by veteran officers who have come to accept the blue code as a normal condition of the police fraternity. Violation of the code results in being "frozen out," distrust by other officers, even failure of other officers to render assistance on calls or to slowly respond when the code breaker needs help.

The Chicago police (among others) have begun drafting plans and policies to begin breaking down the blue code of silence barrier. "The [Chicago] draft rules hold that an officer must intervene if another officer is wrongly using force. Though the department already has a rule against lying, the draft rules specify that officers must truthfully describe uses of force."[12]

Police Personality—Dealing with Compliance Issues

Studies have determined that among the most stressful, and perhaps dangerous, times in a police officer's career are when responding to a "domestic violence" call and when approaching a vehicle for what has been deemed "a routine traffic stop." As officers approach the scene of a domestic violence incident, they have no idea what lies beyond the doorway. What they will normally find are emotionally charged people who are not thinking rationally and have high anger arousal. Any minute the unexpected should be expected, when one of the domestic partners has a weapon and redirects his anger onto the officers.

Non-compliance

The most critical and probable time of danger in a police officer's contact with a potential threat is the initial approach to: the stopped vehicle; a domestic violence incident; a suspicious person or group standing on the sidewalk; or a mentally disturbed individual. The most fearful and frightening time for an officer is conceivably the moment when the contact refuses or fails to comply to the officer's instruction. The mere appearance of an officer in uniform should be sufficient for the offender to know and understand he or she is being instructed by an authority figure and must comply with the requirements. When a person fails to comply, the burden is then on the officer to fully assess the situation and take appropriate and reasonable action to force compliance.

> You're dealing with an erratic, uncooperative subject who outweighs you by 60 pounds, and who has refused to follow your lawful directives to him. Against your repeated directives not to, he reaches into his car anyway. He quickly grabs something from inside and turns with it in his hand. You have thousandths of a second to make a life or death decision. Does he have a weapon? Is it your life, or his? Do you sacrifice your life (and your children's future) for a person who is acting suspiciously, who has failed to follow every one of your legal orders and who may have criminal intent to kill you? Ask yourself: "What would I do in this situation?"[13]

Fear originates in the officer's self-confidence, or lack thereof. Any force beyond "reasonable and necessary" is always considered excess. The officer must immediately take some action, but the extent must be in proportion to the circumstances. The instant an officer puts his or her hand on another person, a fight is likely to ensue. Therein lies the possibility of fear in the officer. A likelihood exists that the offender is better trained in combative techniques and physical conditioning.

There is a clear difference between fear and a threat. Non-compliance alone doesn't constitute a threat, however it should arouse suspicion. People fail to understand how quickly something can escalate and how an officer can become "fearful for his/her life" even when not actively threatened. Just the situation and circumstances can make an officer feel threatened.

If any type of physical altercation takes place and a firearm is pulled, the officer is going to claim he was in fear for his life from an engagement he created when no crime, except possibly a minor misdemeanor, was committed. If the people in the car have committed no crime then any stop feels like harassment. If they ran a stop sign then the only one who should have to produce ID is the driver. No need to engage the others in the vehicle. Many others are becoming familiar with their rights also. The person's race or age should not impact the stop but unfortunately bias often does.

The number one reason officers often increase the use of questionable force is often fear, or they place themselves in tactically disadvantaged situations. Very possibly the officer feels intimidated by the fact, in one instance, the occupants of a vehicle she stops are young black males and in another case the occupants are young gang members. Which leaves the question of what would be the treatment of the young white professionals? Would it be the same and would the officer feel threatened?

If there is reasonable suspicion, does it make sense for a solo officer to attempt to go one on four when no one is in danger? In this situation the officer should radio and call for back up to avoid a potentially tactically disadvantaged position that may force him into recovery, which often leads to elevated use of force.

Is it possible an officer's ego has a lot to do with escalation if a person is non-compliant? Police are human like anyone else. Bias often drives ego. However, no one wants to feel harassed or berated. If facing non-compliance, an officer may feel fear that he may have to resort to physical force, which could be inadequate, so he resorts to higher levels or he may feel powerless leading to frustration and anger, thus over-reaction.

When police officers encounter a person who is non-compliant or displaying a gun, it is imperative that the officers gain control of the situation. This requires that the civilian immediately relinquish his gun or follow the officers' instructions when ordered to do so. Even if the gun is not pointed at an officer, the gun must be given up without further discussion. If the officers tolerate refusal or back talk, they are at risk that the person will swing the gun round and shoot them before they can react.

A "routine traffic stop" is never "routine" because the officer has no way of knowing whom the vehicle occupants are and what they may have done prior to the traffic stop. The occupants may be a carload of retired nuns returning from visiting a fellow nun in a retirement home or they could be gangsters who have recently killed a rival in a drive-by shooting, or any other of an imaginable crime, and will have no hesitation in shooting the officer.

One former law enforcement professional wrote the following public letter:

> I took a complaint about an officer today. The lady that called stated that the officer was rude and in fact "an idiot." She went on to state how unprofessional he was and was extremely condescending towards him in her complaint. I assured her I would look into it and let her know the outcome.
>
> After viewing the recorded stop and speaking with the officer, I had her meet me in my office for a face-to-face. The recording showed the vehicle being pulled over for a headlight out. It was approximately 1am. When the officer gets to the driver's side and begins to greet the driver, he immediately jumps back startled as we hear the driver screaming at him and thrusting something towards him. The officer immediately regains his composure and tries to speak to the driver again. The entire time the driver is yelling "I f*&ckin know!" while he continues to thrust what turns out to be two written warnings from other Police Dept's that night at the officer's face. It goes on for a bit but the Readers Digest version here is that the officer asks the driver to step out and speaks to him on the side of the road (all the while assessing the man for signs of impairment as he speaks and to also be able to hear him over his wife's screaming). He calms the guy down and lets him go with a warning. The officer was professional at all times and never lost his composure. I'm actually amazed that the driver wasn't arrested!
>
> The lady is complaining that they were stopped three times in one night for the same violation. She wants to know how this officer "could be so stupid and pull them over after they were already warned" (in three different jurisdictions across the state). Oh, she also failed to mention that they were driving with their bright lights on and he would not dim them for oncoming traffic so "another moron wouldn't pull them over." I could see I was getting nowhere.
>
> I managed to satisfactorily address her complaint and she left.
>
> The moral of the story is that cops do a dangerous job. Many officers have been killed approaching a car that was stopped for a "minor" violation. Screaming at the officer while making aggressive moves is clearly not appropriate no matter how frustrated you are. That's how a simple stop escalates into something far worse. This lady assumed that the officer should know they had been warned already, and was outraged that he even stopped them. Yet had the officer assumed (and rightly so from the video) that this driver was intent on harming him and took action, she probably would have been outraged at the officer's assumption that he needed to protect himself. After all they are "good people" and not criminals! The officer is supposed to know that. I am dumbfounded how someone can honestly call an officer degrading names and make derogatory remarks for something like this.
>
> Remember, you know that you have been warned already, we don't. You know that you are a good person; we don't when we stop you. We may certainly leave the stop

> with that knowledge, but I assure you that if we start out by you screaming and being aggressive I will have a much harder time believing you! Grandma used to say "you can catch more flies with honey than vinegar." She was probably right.
>
> *XXX Chief of Police, June 12, 2013*

Controversy surrounded the September 19, 2016 non-compliance shooting death of Terence Crutcher in Tulsa, Oklahoma by a Tulsa police officer, Betty Shelby. A call was received by the police dispatcher of a motorist with a stalled car on the highway. One person who called the police department said the driver was "acting strangely and I think he is on something."

The news media quickly picked up the story:

> An unarmed black man killed by a white Oklahoma officer who was responding to a stalled vehicle can be seen in police video walking away from officers and toward his SUV with his hands up before he approaches the driver's side door, where he drops to the ground after being shocked with a stun gun then fatally shot …. Police video shows Crutcher walking toward his SUV that is stopped in the middle of the road. His hands are up and a female officer is following him. As Crutcher approaches the driver's side of the SUV, three male officers walk up and Crutcher appears to lower his hands and place them on the vehicle. The officers surround him, making it harder to see his actions from the dashboard camera's angle.[14]

In an April 2, 2017 interview by CBS's *Sixty Minutes*, correspondent Bill Whitaker talked with Officer Betty Shelby of the Tulsa, Oklahoma police department: In the interview she related that when asked about the disabled vehicle, Crutcher mumbled something she didn't understand and he began to place his hands in his pockets. She ordered him several times to take his hands out of his pockets; he complied and raised the hands over his head.

He lowered his hands and began to place them again into his pockets. Because of her prior experience of a person putting his hands in his pockets and she finding a loaded gun in the person's pockets, she drew her weapon, taking a ready position, and ordered him again to put his hands where she could see them and to get on his knees. He continuing moving back toward his truck.

She said she was thinking, "He's calculating how he can get to his vehicle to get whatever weapon it is that he's going to get because he didn't find it in his pocket …. I'm feeling that his intent is to do me harm and I keep thinking, 'Don't do this. Please don't do this. Don't make this happen.' She also believed when he looked into his SUV and back at her that he was "targeting" where she was so he could retrieve a weapon and shoot.[15] It would have been a good safe practice if the officer believed she was being "targeted" she should have moved, and shifted where she was standing.

Remarks made by law enforcement friends of the author on social media regarding the Terence Crutcher Shooting:

> How does a disabled vehicle response escalate to deadly force? When my car breaks down the thing I hope to see is an officer who may be able to assist. Another dead man, another officer claiming they were in fear of their safety.
>
> When the force continuum goes from verbal commands (even though I didn't hear any, but that does not mean none were given) directly to deadly force considerations the outcomes can be pretty predictable particularly if things are exacerbated by fear as opposed to danger. Always try to separate fear from danger because it is danger that puts

> officers at risk, but it is often fear that causes the public to be severely injured or killed. Every time a person experiences fear does not mean they are in danger.
>
> I have read reports that PCP was found in the vehicle but that information had no bearing on the decision to use deadly force. Because it was not known at the time of engagement. I still don't feel prior to the shooting Mr. Crutcher presented enough of a threat to warrant using deadly force.

These are some very simple questions *prior* to the shooting:

1. Was Mr. Crutcher suspected of committing a crime?
2. Was he under arrest?
3. At what point did he put the officers' lives at risk of severe bodily harm or death?
4. What does a "bad dude" look like (helicopter officer's words in and of themselves are biased)?
5. What made the officer feel the need to draw a weapon on a man in the first place with a disabled vehicle moving away from her?
6. If at any point the officer felt concern for her safety or that of the other officers, couldn't Mr. Crutcher have been tased prior to ever getting in proximity of his vehicle? His back was turned as he walked away so didn't the officer have the element of surprise in her favor? Isn't that the purpose of the less lethal options to gain compliance in a non-deadly force situation?

While these are examples of what perhaps can be fear-causing for their potential instances of death or injury to the officer, what she cannot be prepared for (perhaps not even trained for) is the mental fear of not being able to control an offender who won't comply with the officer's instruction. They don't want to give an inch of the power they have, nor do they want to share the power to dominate. Power of police domination is threatened when police feel "non-compliance" of the offending person they are attempting to take into custody. The fear of "losing" creates, in the mind of an officer, that she must accelerate up the scale of force necessary to maintain dominance of the situation. If fear is allowed to rule one's life, it leads to negative and sometimes violent behavior.

Officers are trained to "use a commanding voice" when giving orders to a subject who is holding a gun or has been fleeing, maybe being subject to a search. The voice of command is thought to intimidate or convince the person to comply with the officer. Because of the high anxiety in such a situation, the officer's voice may reach a pitch that tells the offender the officer is fearful her commands won't be obeyed and forceful compliance must be rendered. At that moment many things go through an officer's thinking process.

The first thing to consider is the adrenalin pumping through both the aggressor and the officer. Each is at heightened alert for movement by the other. The officer could be in a life or death situation and fearful for her life and safety. At what point does she decide to shoot the other person? It is not easy to take a human life. An officer must make a one thousandths of a second decision. In that one thousandths of a second she considers the consequences of her action, the after-shooting investigation and critique, the questions, the news media, the reaction of her co-workers, her friends and family and her own self-doubt. In addition to being in fear for her life she is also fearful of the consequences.

A prime real-life example is that of a Chicago female police officer hospitalized after she was severely beaten. She told investigators she was afraid to use her gun because of the scrutiny from the press and public, plus the legal and departmental consequences she would have faced.

> Police were responding to an auto accident at Roosevelt and Cicero Streets on the city's West Side where officers encountered a violent man under the influence of drugs. Three officers were hospitalized in the incident. One of the officers, a female, was severely beaten and was afraid for her life but afraid to use lethal force with all of the attention on the police department's previous actions and fatal incidents. "She thought she was going to die. She knew that she should shoot this guy, but she chose not to, because she didn't want her family or the department to have to go through the scrutiny the next day on national news," [the] Police Superintendent told reporters.[16]

Failure to comply with an officer doesn't necessarily always lead to use of lethal force, but police are authorized to use "reasonable and necessary force" to acquire compliance. Officers are trained in combatives, self-defense, "takedowns," "come-a-longs," and searching and handcuffing procedures. But, still, an officer must decide when to take physical action and how to subdue the suspect. The mind of the officer may experience anxiety and doubt about his capability to render the suspect under control. What is his first move? Is the combative person more acute in physical fighting? Is he martial arts trained? Is he stronger and more physically fit? At the same time, the officer is being swamped with adrenalin and physiological overloads where he is reacting, not with reasoning but with reliance on his training. It is said by trained martial artists and others who have experienced combat, "You learn a thousand methods, forms, and movements, but when the actual fight begins you fall back on your primary instincts of survival."

In most instances of "police over-reaction" it is because of the fall back on natural instincts to soundly defeat and subdue the opponent. Thus the experience of unreasonable and unnecessary force such as over tasing or beating with a baton. The officer punishes the subject of forcing him to react with violent behavior and causing him to fear being himself beaten. Not only is the officer fearful for his well-being but also of the "hits" his reputation will take from other officers if he does not show he is "manly" and aggressive enough.

If two aggressive people are standing face to face in a confrontation, who strikes the first blow? If you have never been hit squarely in the face or if you have never hit someone the same, would you wait until the other person begins the fight? Once a blow is thrown there is no turning back. Now look at it from the perspective of an officer and the non-compliant individual facing off. What does the officer do?

The elephant in the room of these guidelines is that "excessive force" always ... *always* is only one step above reasonable and necessary. In the heat of the moment or, as some would say, "the heat of passion" when primal instincts, boosted by anger and/or fear, all restraint, regardless of experience and training, dissolves into furious attacking with no thought of eventual outcomes and consequences. When may an officer use excessive force? By definition, the answer is NEVER! Excessive force is never justifiable. Imagine an attorney asking an officer on the stand in a civil law suit case, "At what point did you believe the use of excessive force was necessary?"

The ability to control a situation through communication rather than physical force is *always* preferred. Granted, in some situations physical force will be necessary, but in most situations the ability to de-escalate the situation by problem-solving and critical thinking is favored. Soft skills are proving to be the difference between positive and negative encounters, especially when empathy, active listening, and conflict resolution are embraced. As an ingredient of soft skills, an officer should remember it is not always what is said, but how it is said.

Depending on the circumstance, a responding officer may arrive on a scene where the situation has already escalated to the point of needing to use physical or lethal force. When this happens the assessment and perception in the eyes and mind of the officer are critically important.

Returning to the Tamir Rice shooting previously mentioned; In Cleveland, Ohio on November 22, 2014 a police patrol received a dispatch to a city park concerning a person with a gun. The person calling the police 911 dispatcher said "There is a guy in the park pointing a weapon at people. He keeps putting it in his pants and pulling it out and pointing it. I think it is a fake gun." The part about the gun being a fake was never relayed to the responding officers. Thus was planted in the officers' minds the thought that they were dealing with a potential life or death situation. The officers drove over the park's snow-covered grass to a gazebo where Tamir Rice was sitting with the gun in hand. The police car stopped within inches of Rice and the officers quickly exited. Within two seconds of the officer's arrival, 12-year-old Rice was shot and died the next day.

The shooting became controversial for many reasons: (1) Complete information about the caller's thought that "the gun is probably fake" was never forwarded to the responding officers; (2) The caller estimated the age of person with the gun as about 20 years old; (3) Officers saw the gun and immediately shot Rice; (4) Officers could not explain how or why they felt a life was in immediate danger; (5) What should not be relevant, but adding to the controversy, was the fact Rice was black, which perhaps could have placed a subliminal stereotypical racist fear in the officers' thinking process.

Summary

Immersion into a police subculture begins upon acceptance at a police training academy. In addition to learning police tactics, practices, and procedures, the new recruits, who were hired based on their intelligence, integrity, interpersonal skills, and imagination begin learning a new language (police jargon), adopting a "family," learning unwritten laws (a wall of blue silence—one officer never "rats" or "cops" on another officer, "the attitude test"—a person gets a citation or arrested on the pass or fail score of the attitude of the contact—a form of police use of their discretion), the five steps in the escalation of force—from friendly discussion to use of lethal weapons, and similar topics such as the police code of conduct (an ethical code).

Like any other occupation, officers go through at least five phases during their career. The phases begin with enthusiasm and transit through disillusionment and boredom to waiting for retirement with a pension.

Domestic violence and non-compliance by a contact are two of the many critical times an officer must rely on his or her intelligence and coping abilities to suppress fear and lack of confidence in his or her ability to resolve the situation without resorting to use of force.

Review Questions

1. Examine the role and effects of immersion into a police subculture.
2. Define and discuss the four I's of police recruiting.
3. Define and discuss the five phases of a career in law enforcement.
4. Draw a diagram of the five steps of the escalation of force.
5. When is excessive force justified?
6. Why are examples of domestic violence and non-compliance instances with a high potential for violence and perhaps cause of fear for the officer?

7. Explain the statement, "It is not always what is said, but how it is said that is important."
8. You are a police officer and you stop a car containing four men, ages 20–25 You ask for their identification but they all refuse and are non-compliant to all your orders. They become verbally antagonistic but have committed no crime except the driver drove through a "Stop" sign. It is two in the morning. Emotions are running high. Same scenario but only one person in the car and he refuses all instruction. Any thoughts?
9. How would you deal with these three instances:
   1. They are all black males.
   2. They are white males but have gang tattoos.
   3. They are white males, well-dressed and professional.

Notes

1 Bruno, Benson. *A Story That Talks about Talking Is Like Chatter to Chattering Teeth, and Every Set of Dentures Can Attest to the Fact That No …* www.goodreads.com/author/quotes/5076000.Benson_Bruno. Retrieved March 10, 2019.
2 Williams, Christopher R. and Bruce A. Arrigo A.; *Ethics, Crime, and Criminal Justice*; Pearson/Prentice-Hall Upper Saddle River, NJ 2008; p. 41.
3 www.businessdictionary.com/definition/institutionalization.html. Retrieved January 13, 2019.
4 Goldstein, Herman; *Problem-Oriented Policing;* McGraw-Hill, NY; 1990 as quoted in *Community Policing: Partnerships for Problem, Solving; 7th edn* by Miller, Hess and Orthmann; Cengage Learning, 2014.
5 Skolnick, J. (1977). "A Sketch of the Policeman's 'Working Personality.'" Balch, R. (1977). "The Police Personality: Fact or Fiction." In "Police Personality: What Is It and Why Are They Like That?"; In *Journal of Police and Criminal Psychology* 20(1): 56–67, March 2005; www.researchgate.net/publication/226873077_Police_personality_What_is_it_and_why_are_they_like_that. Retrieved January 1, 2017.
6 The Foundation for Critical Thinking; www.criticalthinking.org. Retrieved December 27, 2018.
7 June, Dale, L. Academic essay.
8 Pittaro, Michael, Dr.; *The Importance of Soft Skills for Criminal Justice Professionals;* https://inpublicsafety.com/2018/09/the-importance-of-soft-skills-for-criminal-justice-professionals. Retrieved October 4, 2018.
9 Adler, Freda, Gerhard Mueller and William Laufer; *Criminology and the Criminal Justice System; 6th edn;* McGraw Hill, NY; 2009.
10 Lawrence Sherman, "Learning Police Ethics", *Criminal Justice Ethics* (1982) appearing in Williams, Christopher R. and Bruce A. Arrigo; *Ethics, Crime, and Criminal Justice*; Pearson/Prentice-Hall Upper Saddle River, NJ, 2008; pp. 40–41.
11 www.navy-net.co.uk/community/threads/the-four-stages-of-policing.41038/. Retrieved April 3, 2017.
12 www.chicagotribune.com/news/local/breaking/ct-chicago-police-use-of-force-met-20161007-story.html.
13 Question in an author survey (D.L. June 2016).
14 http://pittsburgh.cbslocal.com/2016/09/19/police-unarmed-oklahoma-man-had-hands-up-when-fatally-shot-by-cop/. Retrieved January 16, 2017.
15 www.cbsnews.com/news/terence-crutcher-unarmed-black-man-shooting-60-minutes/. Retrieved April 8, 2017.
16 http://abc7chicago.com/news/chicago-cop-says-she-feared-using-gun-while-being-beaten/1543015/.

15

RACISM OR FEAR?

> Being a blackman, a father, son, brother, nephew and cousin, what's stopping me or one of my three brothers, any one of my two sons from getting shot by some police officer who feels we are a threat just because we are black. What would be the thoughts or assumption that I did something wrong? I am a former officer. "I was in fear for my life and my intention was to stop the threat" is what officers are taught to say. So, yes there is specific language that is used when trying to justify a shoot.
>
> *Anonymous*

Name Calling—Verbal Abuse

> So last night some white bum called me a nigger and a spick all in one sentence on Los Olas Street while walking past him. Even told me to go back to my country .... I stopped for a second and right before I hit him in the mouth and borrowed some teeth, I realized, this is how they win .... But I could tell he uses drugs, was homeless, and his anger wasn't really toward me. He wasn't physically threatening so I walked away before turning around to follow him, to turn back around after a few steps, and go about my business .... Else I would have thrown away everything I've been building! ... I have zero tolerance for racism.
>
> *Friend of the Author*

Another friend of the author replied to the above mutual friend:

> Events like this have always been growing moments for me. I think about how fortunate I am, many people that spit hate like this are so angry inside. Life has not gone favorably for them and they don't know what to do, so they lash out at others, anything to make other people as angry or as hurt as they feel inside. That's my point of view on the matter. Life is too good for you to throw it all away for someone that is lost. Your family needs you, they don't care what you have been called at the end of the day, they respect and love you regardless. Those loved ones' words and opinions have been the only ones that matter to me.

Stereotyping

Cultural profiling and stereotyping are generalizations and assumptions made about a certain group of people. Stereotypes can have positive or negative connotations. Nearly all people in society profile and stereotype others to some extent, even if unconsciously. Upon meeting a new person, the first two characteristics we notice about them are their sex and their race or skin color. Racial stereotypes are automatic and exaggerated mental pictures that we hold about people of a certain race. When we stereotype based upon race, we don't take into consideration individual differences. Stereotyping ignores the humanity and uniqueness of each person.

Racial stereotyping is demeaning, devaluing, limiting, and hurtful to the person being stereotyped. It can foster feelings of hate, fear, and aggression that might lead to a false sense of entitlement and superiority. Racial profiling has been a major recruiting tool of white supremacist groups such as the Ku Klux Klan. "Any neurotic drive for supremacy is founded upon a deep and pervading sense of inadequacy and inferiority" (Frances Cress Welsing).

Researchers have documented people's explicit knowledge of the societal stereotypes about blacks in particular. In the United States, for example,

> Blacks are construed as violent, threatening, criminal, un-intelligent, uneducated, lazy, poor, athletic, and musical. People can very easily list the stereotypes of Black Americans, and because these stereotypes are so strong and well rehearsed, they come to influence perception and behavior—even when people do not personally endorse them and are motivated to be racially egalitarian.[1]

Stereotypes linger and become elements in society comparable to urban legends; based on long-time beliefs rooted in half truths, rumors, and exaggerated imaginations propagated by people having experienced exposure to ultra conservative dogmatic values and viewpoints.

A stereotype is an assumption of what people are like, whether true or false. Stereotyping, "when applied to race, is the overgeneralized belief that a certain trait, behavior, or attitude characterizes all members of some identifiable group."[2] Some other current racial stereotypes include the belief that African-American men are "big, muscular, athletic and dangerous"; Latinos are "illegal immigrant 'Mexican' drug smugglers, rapists, criminals … and dangerous." Terrorists are Muslims or Muslims are terrorists, killing and condemning Christians … Muslims want to destroy American and replace Christian values with Islamic beliefs and Sharia law.

Stereotyping derives from/evolves into prejudice, discrimination, and racism. In circumstances involving a police action, fear may be portrayed as a consequence of prejudice, discrimination, and racism. An officer may encounter a person fitting his stereotypical belief and over-react through fear and/or unbridled hostility and aggression.

Racial Profiling

In any culture or subculture there is an underlying current that a person, who in some manner is outside the accepted norm or is "different"… is a deviant. He becomes the one who gets the most attention in an unfavorable way. To be deviant is not criminal, but to be a criminal is deviant (acting outside the accepted legal norm). The label of deviance is based on wrongful stereotypes.

Consider a scenario of a white male walking in a predominately black neighborhood. A police patrol car drives by and the white pedestrian waves at the officers in the car who return the wave and continue on their way. In a similar situation, a black male is walking in a primarily white neighborhood. It is irrelevant if the pedestrian waves at the officers or not. The officers will direct their attention to him and stop and question him about why is he in the area, what is his destination, what is his intent, and who does he know in the neighborhood. He will be required to produce identification and wait while the officers, one with his hand on the butt of his weapon and the holster unsnapped, the other holding the identification, conduct warrant searches. The officers will "pat down" the person, who will be referred to as "suspect," for weapons before allowing him to proceed on his way. The "suspect" may have to empty his pockets or lie spread eagle on the ground while officers conduct their "investigation."

Same scenario, different approaches; based on nothing more than a person's skin color and stereotype or "racial profile." Racial profiling is

> when law enforcement and private security target people of color for humiliating and often frightening detentions, interrogations, and searches without evidence of criminal activity and based on perceived race, ethnicity, national origin, or religion. Racial profiling is patently illegal, violating the U.S. Constitution's core promises of equal protection under the law to all and freedom from unreasonable searches and seizures. Just as importantly, racial profiling is ineffective. It alienates communities from law enforcement, hinders community policing efforts, and causes law enforcement to lose credibility and trust among the people they are sworn to protect and serve.[3]

Is It Racism or Is It Fear?

Fear is a catalyst for racism. As a catalyst, fear accelerates the imagination. Fear fosters resentment; resentment fuels prejudice. Prejudice is a form of racism as it relates to treatment of those deemed "inferior" and leads to oppression from the dominant people with the most power.

> Whites are generally not followed around in stores or asked for identification when no one else is questioned. They are not passed over by taxi drivers or subjected to constant media images of people who look like them in handcuffs. Whites do not have to see people react to them by clutching their purses and crossing the street and are not regularly subjected to police brutality .... If many whites feel anxiety in today's America, many blacks feel an existential threat that seems never to end.[4]

As the first element in administering justice, law enforcement understands the obligation that justice means treating everyone without prejudice or bias but receiving what is due, whether it is punishment or reward. Ethical fairness demands no tolerance for discrimination based on race, religion, sex, or age.

Altruism is very common in the people who choose a career in criminal justice. Doing good only for the sake of goodness itself is endemic, or should be, in the personality of the officer. His personal motto should be "We are here to help." In any given circumstance the power to help motivates an officer to do his duty without thought of personal gain. The thought of not assisting another is an abomination to his personal moral code. Beneficence means doing good merely for the fact of doing the right thing for their own sake. The obligation of non-maleficence means doing no intentional harm or causing injury to others. A criminal justice practitioner following the principles of good ethics and moral behavior will follow a path of refraining from using force or lethal methods unless it is "reasonable and necessary."

In a 2018 criminal justice class "psycho drama" assignment of students acting out scenarios of their vision of a police stop of a young black man, one black student "brought the house" down with his interpretation of complying with an officer. First he gave his impression from the white officer's view:

> Say, Boy were you going too fast? Let me see those hands on the steering wheel! Do you understand me? One hand out the window and the other open the door and get out slowly. I don't want to hear any disrespect from you and you better comply or you'll be face down on the ground and you'll be doing hard time.

Switching to a southern slave style accent and dialect, the student played the part of a young black male as he thought the police wanted the driver to react.

> Yes, suh, I musta be goin' too fast ... Yes, suh mistah pooo lice mans, I be havin' my hand on the wheel ... Yes, suh, yes, suh, I knows what you say. ... I be doin' it now ... I ain't got no gun 'n no other stuff ... If you says so I's gittin' slow outa de car ... I don't mean no dis 'spect, mistah suh .... I's 'plyin' as you say.

That was done, of course, for laughs as the student portrayed the scenario in words and pantomime; the other black students in the classroom laughed and clapped their hands while nodding their heads in agreement.

Following the presentation, a white student gave his interpretation, first as the white officer and then as the black driver.

> Good morning, Sir, I see you were going a little over the speed limit. This is a school zone and there are lots of kids around and one may dash into the street. May I please see your driver's license, car registration and insurance card ... If you don't have them I must give you a citation to appear in court ...Would you please step out of the car and over on the side of the street so we won't get hit by passing cars?

Next the student became the black driver.

> Why'd you stop me? ... I aint doing nuthin' ... I'm on the way to see my parole officer .... I don't need to be stepping out of the car ... You stopped me cuz I'm black, you're profiling, and you think I might have some smack in the car or you need to meet a quota and I'm your boy today ... Well, I'm not anybody's boy ... My license is in my wallet in the glove box ... I'll just reach in and get it ...

The officer pulls his weapon and shoots the driver, "because he made an unexpected move and I was afraid for my life."

In the third scenario a student played the dual role of a white police officer and a white male driver.

> Good morning, Sir. I stopped you because you were driving forty miles an hour in a fifteen miles per hour school zone. Are you late to work or an important appointment?
>
> Yes, Officer, and good morning to you also. I have a meeting with a prospective client in downtown that could mean a big financial gain. I'm sorry I didn't realize I was going over the speed limit. I won't do it again.
>
> Well I patrol this area every day, so I'll let you go this time; just be more aware of the speed limit. I'd hate to see a young kid run out and get hit.

Thank you, Officer.

Over exaggerations, and stereotypes but to the students in the class it was how they see and interpret their reality.

White rap artist, Eminem, in his song, "Untouchable," addresses issues between white police officers and black males by using the words, "white boy" as if black males are talking to white officers. He then uses "black boy" to make it appear that white officers are talking to black males to create what appears as a dialogue. He raps about white officers hating black males and profiling them.

In the lyrics, Eminem explains how white police officers do not like the sight of black males and how officers pull up on them and stereotype profile them. Then he adds the line about black males coming from a "dope house" because when white officers profile black males, they make assumptions based on stereotypes and comparisons.

The song shows the mindset of a racist officer intimidating a black man he has stopped for a minor vehicle violation. According to black males, this way of thinking and speaking has been a standard procedure during interactions with white police officers. It is a racist fact in the speech of the officer and fear in the mind of the black male.

Eminem switches his song to a black voice and mentions how white officers, or white people, in general are considered to be "untouchable"; From this Eminem is illustrating how white officers are "untouchable" because they are police officers in their police cars, usually in cases regarding a white officer taking extra measures to take down a black suspect, the officer gets little to no punishment.

In another verse of the song, Eminem highlights how white officers are ignorant of black culture and how officers do not care for the excuses black males make as to why they are victims in the world. Eninem makes a point that white officers do not care about what excuses or alibis black males have when it comes to why they committed or are suspected of a crime.

He continues by saying how white police officers view black males and how they do not see their lives as being meaningful. This verse also implies how white officers believe they are the masters in charge and anything they say is the way things are "going down." If black males fail to listen (or comply) they will get shot.

Of course Eminem is a white musical artist but his experiences and thinking mirror those of a black man in today's black/white uninformed society.

A viral post on YouTube showed a white police officer in Georgia stopping a white woman. She immediately put her hands up and told the officer she was afraid she would get shot if she put them down to get her phone from her lap. The officer replied that she wouldn't get shot because she was white and they shoot only blacks.

To give the officer a benefit of doubt perhaps he was being sarcastic and thought the remark was funny, while in reality it was poor judgment, racist, and hurtful to all blacks. One reply by a black male was, "When that's how you feel, it's just a regular day to them."[5]

A 2016 study by Holbrook, Fessler, and Navarrete hypothesized that:

> both formidability and status are conceptualized in terms of bodily size/strength, and that formidability representation will take precedence over status representation when threat is salient. This hypothesis generates four predictions:
>
> - Individuals belonging to groups stereotyped as threateningly violent will be envisioned as more physically formidable—despite being regarded as lower in status.
> - Differences in envisioned physical formidability will mediate differences in attributions of physical aggressiveness (i.e., propensity for violence).

- Envisioned physical formidability and status will positively correlate in judgments of target individuals whose group is stereotyped as non-threatening, but not in judgments of target individuals whose group is stereotyped as threatening.
- Controlling for group differences in perceived aggressiveness will reduce or eliminate differences in the correlation between physical formidability and status.

They concluded that:

> prevailing impressions of Black and Hispanic men as large and muscular [modern sports media have created a stereotype of black and Hispanic men as larger and stronger than white men—though demographics have shown the average black and Hispanic man are equal in size to the average white man at five feet nine inches and 190 pounds] are connected to perceptions of physical aggressiveness, and that associations with danger are detectable via patterned differences in the conceptual links between physical size, status, and aggressiveness.[6]

Given all the data above, combining it with an officer's perception of an emergency occurrence involving a black man, and given information that a gun may be involved, with adrenalin pumping from racing to the scene, perhaps with red lights flashing and siren screaming, the officer may (or probably will) be induced to resort to physical or lethal violence.

> Further, as policing is a dangerous occupation, an officer's fear for his life shapes his conduct. Van der Wolf (1962), argues that the probability of threat and the formulation of a response to it place an officer at a disadvantage in a citizen encounter. Similarly, Robinson (1987) argues that, in part, brutality is the result of fear and consequent overreaction based on perceptions to potentially dangerous situations.[7]

The division between law enforcement and citizens has been perpetuated by the ignorant belief that law enforcement is systematically targeting minorities. Although there have been instances and occasions when unethical (perhaps illegal) law enforcement practices have made headlines, such practices are not the norm. Furthermore, such practices are so deplorable that law enforcement officers who see such acts will take necessary actions to overcome and report illegal policing practices. Contrary to news and social media reports and perpetuation, law enforcement officers go to work daily to protect the citizens they have sworn to protect from the evils of the world and ensure freedom from victimization regardless of their race, sex, sexual orientation, religious or political beliefs. When law enforcement officers look at citizens, they see a chance to preserve the good in humanity, not an opportunity to undermine it.

The Color of Fear

Psychologists have researched the effects of color on people and concluded that color has profound effects. Color has a psychological effect on the minds of humans as it relates to clothing, packaging, paint on a house and rooms, even jail cells (painted pink as a color of tranquility). What is the color of fear? Does yellow mean fear?

> The right yellow will lift our spirits and our self-esteem; it is the color of confidence and optimism. Too much of it, or the wrong tone in relation to the other tones in a color scheme, can cause self-esteem to plummet, giving rise to fear and anxiety. Our "yellow streak" can surface.[8]

"You dirty rotten yellow-bellied coward," often heard in a western movie, is a challenge thrown by a bully at his target. If the subject of the bully's fury becomes frightened, we can assume the color of fear is yellow. Other quotes, "He has a yellow streak down his back" and "He's yellow," also assume someone is a coward or frightened easily. Alluding "yellow" to fear and cowardice has no one historically agreed upon origin. An online search for the term "yellow belly" showed several versions. The one recurring theme is that "yellow" referring to a person means coward.

If yellow means coward, what of the color black? "Black is associated with power, elegance, formality, death, evil, and mystery. Black is a mysterious color associated with fear and the unknown (black holes). It usually has a negative connotation (blacklist, black humor, 'black death')."[9] At Halloween the most favored color of the costume is black, as in witches, vampires, and other frightening creatures. In the past, executioners wore a black hood; funeral undertakers wore black and birds of carrion are black.

In a 1997 study by Hupka et al., "661 undergraduates indicated on 6-point scales to what extent anger, envy, fear, and jealousy reminded them of 12 terms of color. In all nations, the colors of anger were black and red, fear was black, and jealousy was red."[10]

What of the effect of the color of a person's skin, the texture of his hair, his culture, his physical environment? Historically, especially in the United States, the color of a white person's skin is equivalent to prestige and privilege; a black person's to a status of threatening, second-class citizen. This type of attitude has often been observed by treating a dark-skinned person as "subhuman" and dangerous.

> Recent shootings of black men by white police officers have highlighted a deep, uncomfortable truth about many Americans: *They fear black men.* The stereotype of the black man as the aggressor, a physically powerful, unpredictable product of poverty, has held sway since the Great Migration saw southern African-American communities relocate to northern cities, triggering "white flight." That's obviously a problem, *full stop*, but it's a particularly intractable problem in stressful situations according to Colin Holbrook, who researches behavior, evolution, and culture at University of California- Merced. Holbrook focuses on decision-making under perceived threat. His research shows an inherent distrust and fear of a black man. According to his studies, a white man perceives a black man as large and powerful.
>
> Holbrook and his team found that when a white character was described as physically large, participants imagined him to be prestigious and respectable, commanding a room and owning it. But change one descriptor—add "black" to the list of traits—and toss on a culturally stereotypical name and everything changes: Participants imagine someone potentially dangerous, probably poor, maybe criminal.[11]

> Cops can get into a state of mind where they're scared to death. When they're in that really, really frightened place they panic and they act out on that panic. I have known cops who haven't had a racist bone in their bodies and in fact had adopted black children, they went to black churches on the weekend, and these are the white cops. But you know what they had in their minds that made them act out and beat a black suspect unwarrentedly? They had fear. They were afraid of black men. I know a lot of white cops who have told me. And I interviewed over 900 police officers in 18 months and they started talking to me. It was almost like a therapy session for them.
>
> They would say things like, "I'm scared of black men. Black men terrify me. I'm really scared of them. You know black men who come out of prison, they've got great hulk strength and I'm afraid they're going to kill me. Can you teach me how not to be afraid of black men." I mean these [are] cops who are 6'4". You know, the cop in

> Ferguson was 6'4" talking about he was terrified. But when cops are scared, they kill and they do things that don't make sense to you and me.[12]

> In a diverse country like the United States, race complicates everyone's story. Steven Neuberg (an experimental psychologist) says that in general, young men tend to have stereotypes of being more criminal, more impulsive, and more dangerous. The narratives we imagine for black men are particularly bleak. The problem, in other words, is both the reality born of historical prejudice and injustice, and the fantasy, born of a natural impulse and subconscious prejudice.
>
> It's this divide that creates a sense of enmity between young black men and cops, one that's become increasingly stark in the past few years. "You see why police are wary, why people of color are wary," Neuberg says. "From both perspectives, the other one is considered dangerous. And so much of that comes from thoughts each group has about the other group's background: for young black males, the fact that white cops see themselves in a position of power and able to dole out punishment at will and for no apparent reason; for white cops, a sense that young black males are out to stir up trouble. And when you toss in the fact that young black men might be more likely to be economically distressed, it further alienates each group.[13]

Dehumanization

Envisioning a dark-skinned person as large and dominating can be taken as having extraordinary strength and prowess commensurate with a "non-human." Eliminating all feelings of humanity against another makes it easier to treat that person as a non-human object that deserves to be beaten, confined, or killed. "As the scholar of warfare Johannes Lang has observed of the Nazi death camps, 'What might look like the dehumanization of the other is instead a way to exert power over another human.'"[14]

It would be wrong to assign dehumanization to people of color without mentioning the homeless and the drug-addicted. Seeing a homeless person huddled on a bus or park bench or asleep on a sidewalk, most people will simply pass by or step over the person as though they are invisible or simply an object to be avoided. They are avoided because of the stereotype of a homeless or drug-addicted person being mentally ill and dangerous. People in those categories are easily overlooked or avoided because they have become dehumanized in the eyes of the average passerby.

> In his landmark book on metaphorical representations in popular media, Otto Santa Ana (2002) argued that media outlets tacitly compare Mexican immigrants to insects, among other things. Santa Ana argued that in newspapers, this happens when words such as *swarm* or *crawl* are used to describe Mexican immigrants. Santa Ana further argued that these words, linking Mexican immigrants to insects, create an implicit metaphor that, although hidden to casual readers, powerfully impacts the ways in which they conceive of Mexican immigrants and issues surrounding immigration.[15]

The Color of Racism

Is black the absence of color or is white?

> In light, white is the combination of all the different colors of light. Black is the absence of light, therefore the absence of all colors of light. In pigment, white is the absence of color. Black is a color rendered so darkly that the individual color cannot be perceived.
>
> *Yahoo Answers*

Daily news media and the Internet are loaded with stories of police shootings and use of excessive force against black and brown men and boys. Is it a result of over-sensitivity to a potentially dangerous situation seen through the eyes and mind of the officer as he was trained or is it based on a bias and prejudice evolved from his experience? Does a person's skin color and other physical features affect the officer's judgment? Is the officer influenced by stereotypes he has learned and believes?

A new officer begins his career as energetic and believer in equal treatment for all. As time passes, he becomes exposed to every stereotype he has been told about. Over the course of time he grows cynical, distrustful, and firm in his "warrior" mindset and perspective that everyone (especially black men) are to be treated with extremeness, caution, and force.

Since the days of slavery, the propagation of the myth of the predatory Black man has been used to instill fear in whites and to justify their brutality and violence against Black individuals, communities and continents. The narrative has been passed down from one generation to the next and is still used to underwrite injustice against Black people.

The enslavement of Africans resulted in a plethora of uprisings, from the Haitian Rebellion[16] to Nat Turner's rebellion.[17] Since then, whites have developed a pathological fear that oppressed Blacks will one day rise up and inflict vengeance upon their oppressors.

psychological defense mechanism called projection—when one accuses someone of having traits they refuse to acknowledge in themselves—may also explain why white people fear violence from Black people. Instead of acknowledging the past and present forms of violence Black people have suffered at the hands of whites, it is projected onto the victims themselves.

Historically and up until present times, the media continue to broadcast daily imagery of Black men as dangerously criminal, using and dealing drugs, hypersexual, unemployable, idle and the epitome of death and doom. This is true despite the reality that white people have, and do, participate in mob and domestic violence in higher numbers, and that whites comprise more than 70 percent of drug abusers and dealers in our country.

White people's fear of Blacks also stems from a form of moral blindness, in which Black bodies remind them of the terrible crimes from an ugly past, but they are unable to own up to it because they have bought into the Black predator myth and practice projection. Instead, they deny, shift blame, lie, twist facts and characterize Black people into a people to fear. Whites also fear becoming outcasts among whites by associating and sympathizing with Black folks.

Dr. Frances Cress Welsing, author of *The Isis Papers*, [18] argues that because whiteness is genetically recessive, some whites fear that integration and miscegenation will result in the annihilation of the white race. Therefore, they established a global system of "White Supremacy" and its subsystems of racism and segregation to prevent it.

Many whites who directly benefit from the current plight of Black people fear that their enlightenment and subsequent organizing will result in the inability to exploit and benefit from their resources, as well as a reclaiming of their historic legacy.[19]

Source: Moore, A. "8 Reasons Why White People Fear Black People"; September 24, 2014; https://atlantablackstar.com/2014/09/24/8-reasons-why-white-people-fear-black-people/. Retrieved October 17, 2019.

The difficult relationship between the police and blacks in the United States is evident in the deep-rooted racial divisions in the public's view of law enforcement and the criminal justice system.

But the latest Associated Press-NORC Center for Public Affairs Research poll [July 17 to 19, 2015] also finds agreement across racial groups on many of the causes of police violence.

Some of the poll's key findings are:

- Violence against civilians by police officers is an extremely or very serious problem according to nearly three-quarters of blacks and less than 20 percent of whites.
- Many Americans, both blacks and whites, say that violence against police is also an extremely or very serious problem in the United States. *And half of all Americans, regardless of race, say fear caused by the physical danger that police officers face is a major contributor to aggression against civilians.* [20]
- An overwhelming majority of blacks say that, generally, the police are too quick to use deadly force and that they are more likely to use it against a black person. Most whites say police officers typically use deadly force only when necessary and that race is not a factor in decisions to use force.
- Blacks and whites are sharply divided on whether police officers who injure or kill civilians are treated too leniently by prosecutors and on how much that contributes to the use of force against members of the public.
- Half of black Americans report being treated unfairly by police because of their race, and their views of law enforcement are shaped by this experience.[21]

The same survey found:

> A large majority, 81 percent, of blacks say police use deadly force too quickly compared with 61 percent of Hispanics and 33 percent of whites. Similarly, 85 percent of blacks think police are more likely to use force against a black person in most communities, compared with 63 percent of Hispanics and 39 percent of whites. Nearly as many, 71 percent, of blacks say police in their own community are more likely to use force against a black person compared with 47 percent of Hispanics and 24 percent of whites.[22]

"As a black man," (Richard) Jeanty[23] says, "I want people to understand what it feels like to be in my shoes. I know that I am a potential victim." And it makes no difference if you are from the "hood" or live in the suburbs, in your car or in your home, as Professor Gates[24] discovered, and Earl Graves, Jr., another professional African-American man, found out when he was stopped in 1995, and searched by Metro-North Police.

Jeanty adds,

> Make no mistake, the message is loud and clear: no matter who you are in this society, as long as you are black, you can still be stopped, searched and arrested by the police. And all black men know, a seemingly minor interaction with the police can turn deadly.

The following personal story is from "Joseph," a criminal justice student working on his Bachelor Degree. He volunteered his essay exclusively for this book.

> You as society are responsible for allowing Law Enforcement for all these years to act reckless. Many of us are scared from you never believing in our stories and experiences. The first time I was arrested was because I was on the hood of my friend's car, driving down a neighborhood street more or less. I was 18yrs old. They also claimed I resisted arrest with violence (ALL charges were dropped and the judge actually laughed at it). There was no statute on this, and I was essentially handcuffed, arrested and sent to jail for a few days for nothing. Did I mention this was the 2nd time Law enforcement beat the shit out of me, using an entire can of mace on my eyes, burning my scalp and eyelids …. Although it was dropped, the same day I went to court, it remains in my record.
>
> The 2nd time I was arrested I was grabbed up in one round up for loitering. I literally just came out the store and tried to lite my black and mild cigar. As I stroke the match head to the pack, the officer told me to put it out and sit down. I was being arrested for trespassing and loitering. He pulled 4 of us into the back of his car. I didn't know the other guys, but they all knew each other including the officer. There was no reason for this interaction, much less arrest. That was my second time being arrested. A great way to send me down a path where crime came easier and being a return offender was definite.
>
> These are the processes used to hinder people of color across the country. They get you in the system, and then its starts going down hill from there, or what it's worth, I have seen all types of people abused by law enforcement, and because I don't know particulars to others cases, I can't speak on them. I do not think this is a race issue, as much as it is a "policing" issue. But I can say based on any common research, we will see a questionable amount of convictions overturned, as well as many cases of excessive force, never uncovered. What do we do to change this? I don't think we can. Much of it is rooted in culture and the social science of the job. But it's important for the new society to remember there are a lot of men out there that belong to certain communities that didn't have a fair day in court, or were arrested under false accusations and shady means.[25]

These comments and studies could lead one to conclude that racism is the leading cause of unequal treatment between persons of color and whites, and that police all too frequently use unreasonable force, including lethal violence, against more persons of a racial description than whites. There is also evidence from other studies that police use force and lethal weapons, not from racism but from fear and frequency of contact more often with African-Americans and other minorities than whites simply because of social factors such as high crime rate areas and environment, social disorganization, poverty, etc. Study after study, though statistics vary and are often incomplete, show minorities are stopped and interviewed by police more frequently than are whites but the percentages of a violent encounter leading to shootings remain proportionally equal.

Summary

Stereotypes are commonalities, often wrongfully misconstrued from ignorance of one group about another. Stereotypes like "blondes are dumb" are found in every strata of life and some are hilarious or the butt of jokes. However racial stereotyping has a long history of hate, bedded in fear.

Racial stereotypes are, of course, the worst kind of stereotype; but often used in policing tactics of stopping and questioning persons of color. It is simply a jaundiced means of using a stereotype to conduct stops and field interviews and make note of the encounter. The practice of racial profiling is not ethical and is not legal in most states.

Review Questions

1. What is wrong with stereotypes and racial stereotyping for policing purposes?
2. What is the color of fear, or is there no color of fear only a stereotype?
3. Discuss the history of fear of black men by white men.

Notes

1 Goff, Williams, Eberhardt, and Jackson; "Not Yet Human: Implicit Knowledge, Historical Dehumanization, and Contemporary Consequences." *Journal of Personality and Social Psychology*, 94(2), 292–306; Copyright 2008 by the American Psychological Association. Retrieved March 16, 2018.

2 Newman, David, M.; *Sociology, Exploring the Architecture of Everyday Life;* Sage Publications; Thousand Oakes, CA 2010.

3 www.aclu.org/issues/racial-justice/race-and-criminal-justice/racial-profiling.

4 Psychologist Beverly Tatum; Quoted in *Political Tribes: Group Instinct and the Fate of Nations* by Amy Chua; Penguin Press; New York, 2018, p. 175.

5 Sent to me by a very good friend, an African-American, Andre Glenn.

6 Holbrook, Colin, Daniel M.T. Fessler and Carlos David Navarette; "Looming Large in Others' Eyes: Racial Stereotypes Illuminate Dual Adaptations For Representing Threat Versus Prestige as Physical Size"; *Evolution and Human Behavior*; August 28, 2015; http://colinholbrook.com/Holbrook_Fessler_Navarrete_Status_Threat_Size_EHB.pdf; Retrieved March 14, 2018.

7 Lersch, Kim Michelle; *Policing and Misconduct*; Prentice-Hall; Upper Saddle River, NJ; 2002, p. 45.

8 Psychological Properties Of Colours; www.colour-affects.co.uk/psychological-properties-of-colours. Retrieved 03.11/18

9 www.bing.com/search?q=colors+asociated+with+fear&qs=HS&pq=c&sc=8-1&cvid=D4AF38D3D08348ADA0F4E8F828159CFB&FORM=QBLH&sp=1.

10 Hupka, Ralph B., Zbigniew Zaleski, Jurgen Otto, Lucy Reidl, and Nadia V. Tarabrina; "The Colors of Anger, Envy, Fear, and Jealousy: A Cross-Cultural Study." *Journal of Cross-Cultural Psychology*; March 1, 1997; http://journals.sagepub.com/doi/pdf/10.1177/0022022197282002. Retrieved March 12, 2018.

11 *Why Americans See Black Men as a Threat (We Don't Just See Skin)* by Tanya Basu (July 12, 2016) www.inverse.com/article/18159-science-of-racism-against-black-americans-police-shootings-research. Retrieved March 12, 2018.

12 Williams, Joseph, P. *When It Comes to Police Brutality, Fear is Also a Factor*; www.usnews.com/news/blogs/washington-whispers/2014/12/05/when-it-comes-to-police-brutality-fear-is-also-a-factor. December 5, 2014; Retrieved March 21, 2018.

13 Op. cit, Hupka et al.

14 Bloom, Paul. www.newyorker.com/magazine/2017/11/27/the-root-of-all-cruelty. Retrieved March 16, 2018.

15 Op. cit. Holbrook et al.

16 history of the Haitian Revolution can be read at www.blackpast.org/gah/haitian-revolution-1791-1804.

17 The Confession of Nat Turner can be found at http://docsouth.unc.edu/neh/turner/turner.html.

18 http://kentakepage.com/7-things-you-would-have-learned-if-you-read-the-isis-papers/.

19 Moore, A.; "8 Reasons Why White People Fear Black People"; *Atlanta Black Star*; September 24, 2014; http://atlantablackstar.com/2014/09/24/8-reasons-why-white-people-fear-black-people/4/; Retrieved July 9, 2019.

20 Italics by the author for emphasis as that is the main theme of this book.

21 *Law Enforcement and Violence: The Divide between Black and White Americans*; The Associated Press-NORC Center for Public Affairs Research; www.apnorc.org/projects/Pages/HTML%20Reports/law-enforcement-and-violence-the-divide-between-black-and-white-americans0803-9759.aspx. Retrieved March 17, 2018.

22 Ibid.

23 Jeanty, Richard; *Why African-American Men Fear The Police: The Most Dangerous Gang in America: The Police*; www.blackradionetwork.com/why_africanamerican_men_fear_the_police. Retrieved March 17, 2018.

24 Prominent Harvard Professor Henry Louis Gates, Jr. arrested, suspected of burglary as he was attempting to enter his own home, July 16, 2009.

25 Observations and experience written by "Joseph," a criminal justice student working on his Bachelor Degree.

16

REALITY AND THE BRAIN

Reality of the Mind—Conscious or Unconscious?

It's all in the mind of the beholder. If it doesn't exist in your consciousness it doesn't exist. Look around you. The people, cars, trees, buildings, etc. did not exist in your reality until you brought them into your view and your brain told you they existed. Of course they existed but not in your perception. By seeing them they were brought into your conscious reality. Fear is an emotion but it doesn't exist until the senses tell the brain of pending danger. The brain constructs individual realities and imprints what we are accustomed to but adapts to change.

The change the brain makes is not mere adjustment. The change actually is "rewiring" and making of new or altered electron connections, often mistakenly referred to as "muscle memory." As input is made from a new experience, the movement might be very awkward until it is repeated over and over again and the brain becomes accustomed to the newly built pathways. "The eyes can't see what the brain doesn't know."

> A brain is only capable of what it could conceive, and it couldn't conceive what it hasn't experienced.
>
> *Graham Greene*

This also begs another question in a cliché. We know that to make a sound there must be a source and a receptor. Thus the old question, "If a tree falls in the forest and there is no one to hear it, does it make a sound?" Of course it does but not in anyone's reality.

An example of not seeing or recognizing something, even danger, if the mind is concentrating on one thing, leaving everything else out of the conscious mind, is the time two patrol officers were dispatched to the scene of a large street fight in a very violent neighborhood at ten o'clock at night. As they entered the darkened street (several street lamps were off) they immediately believed it was a false call because the street was vacant of anyone. Suddenly several gun shots rang out and their patrol car was struck. The officer in the passenger seat said calmly, "They are shooting at us." The driver, concentrating on looking for a disturbance, replied "Ok," and kept driving at the slow patrol pace. The passenger officer again said, in a more immediate tone, "They're shooting at us!" The driving officer said, "What? You said they're shooting at us?" "Yes, get us out of here!" came the reply. Then and only then did the reality of the situation come into the consciousness of the driver who then immediately raced from the scene.

There is an interesting "semi-psychological" or mental game that can help bring imagined situations and circumstances into your reality. For lack of a better term, the game is "What if ..." It is a game that can be played as a solitary endeavor or with any number of participants. An imaginary but likely or possible scenario is described beginning with the words, "What if ..." As the mental process takes place, the person is solving the problem in advance for if and when he encounters the same scenario. He can and will respond as an automatic reaction because his "mind's eye" (his reality) had already visited similar circumstances thus creating experience, at the mental level, allowing him to respond physically.

Recognizing what is being seen is also a criteria of reality. A naive or untrained person may see two people seeming to innocently meet on a street and perhaps shake hands, exchange a few words, and proceed on with their walk. A police officer may view the same scenario and recognize it as a drug buy. It's all in the eye and mind of the beholder and what her mind is focused on relating to her experience and sense of reality. No two people see the same reality. This is a matter of perception, influenced by many factors (to be discussed later). There have been times in a person's life when his focus is elsewhere when a frightening object or person comes into view. It may take a hesitant second for the person to change his focus and realize he is seeing something very frightening. He then reacts in a corresponding manner of adrenalin flow, rapid heartbeat, freeze or flight, most probably his limbs will become numb, and he will be unable to move. He may even experience a release of bodily waste. It is the surprise, the sudden shock, that freezes him, maybe even when the sudden recognition of what he perceives as frightening is no threat at all.

How the Brain Constructs Our Sense of Reality

The brain constantly receives messages from the senses. It never sleeps but merely rests. Even when the body is asleep the brain continues receiving messages such as telling the body if it is too warm or too cool, if a light comes on or a disturbing sound is detected; the brain then calls for a response. If danger is perceived, the brain sounds an alarm and places the body on alert, and physiological changes prepare the fight, flight, or freeze reflex.

A sensation is felt, a thing is seen, a sound is heard, a food is tasted, and an odor is smelled; each travels through a complex network of trillions of electrical impulses in the brain which interprets the sense and orders the response.

Consciousness—Unconsciousness

The brain controls all the bodily functions; a small part of the control is exerted by the conscious mind, the remaining control is out of our awareness and is processed in the cerebellum. Consciousness can be rudimentarily defined as awareness of all things in the observer's reality by the awakened body and is influential in decisions we make. But it is only a small part of what is happening in the brain. Activity of the unconscious responds without any cognitive thought while the conscious does not register anything and must be stimulated by the senses.

Sigmund Freud said that beneath the surface of us lies a vast expanse of hidden motivations, desires, and drives. The way we think, feel, and act is greatly influenced by the unconscious. Most of our fine decisions are not consciously made but come from unconscious activity. Unconscious action is derived from education, training, and experience. It may be those conscious activities reacting in the unconscious that leads to errant behavior as reactions to the stimulus of training and learned response.

What happens when we encounter unexpected things? The conscious mind operates two tenths to five tenths of a second behind the unconscious, perhaps causing a wrong decision or activity. It may be thought that with the unconscious operating ahead of the conscious that Man's consciousness operates in the past. But when experiencing sudden unexpected things, it is the conscious mind that assesses the situation for a dangerous threat or opportunity and to make sense of it. If a dangerous threat is determined, the conscious sends a message to the brain ordering it to release adrenalin to cope with the situation through the fight, flight, or freeze reflex. That is why a sudden attack or threat momentarily stuns the victim until he can consciously make the proper assessment. It's not uncommon for an attacked person to momentarily ask himself, "Why is this person attacking me, I didn't do anything to him?"

A highly trained police officer, having had the dangerousness of his job drilled into him at his academy and through experience is always suspicious of every encounter and is on edge, ready to react to any movement his conscious brain tells him is dangerous (professional paranoia). Sometimes his unconscious brain detects something not yet in his consciousness and he is suddenly surprised when he consciously becomes aware; thus he may over-react, resulting in action leading to injury or death of the person he has encountered. That may lead him to believe or say, "I felt my life was in danger so I reacted according to my training and department policy." But of course training and departmental policy cannot foresee all circumstances the officer will encounter, leaving him to exercise his professional discretion relying on his conscious and unconscious brain.

A police officer being called to the scene of a reported robbery in progress may experience a range of emotions. On the way to the crime she may initially feel a rush of excitement and as she proceeds she may mentally rehearse her approach and attempt to calm her nerves and breathing. But with flashing red lights, a siren wailing, and driving at high speed, her body is pumping out high doses of adrenalin, bringing her to a sharpened edge of awareness doubled with fear as she anticipates facing the hazards and the unknown.

Upon arrival she sees the robbery suspect running out the door with something in his hand. Of course the officer also has her firearm in hand. She has thousandths of a second to make an accurate assessment. What does the robber have in his hand? Is it a gun? What are the actions of the robber? Is he running away? Does he see the officer? Does he turn toward the officer? Is the officer's life in immediate danger? Are there bystanders in the line of fire? Those are thoughts running through the mind of the officer in less time than it takes to read this sentence. Of course we shouldn't expect the officer to wait for the robber to open fire toward her before deciding to shoot; delay in shooting first can result in her being shot, perhaps killed. At what point does the officer shoot? What if she shoots and then finds out the "something in his hand" was a bottle of soda and the "robber" was not committing a robbery but a theft, simple shop lifting, a "grab and run"?

How Does Fear Affect Human Behavior?

Fear is "an unpleasant and often strong emotion caused by anticipation or awareness of danger."[1] Fear is completely natural and helps people to recognize and respond to dangerous situations and threats. However, healthy fear—or fear which has a protective function—can evolve into unhealthy or pathological fear, which can lead to exaggerated and violent behavior.

Dr. Ivan Kos lays out several different stages of fear. The first is real fear, or fear based on a real situation. If someone or something hurts you, you have a reason to fear it in the future. Second is realistic, or possible, fear. This is fear based on reality that causes a person to avoid a threat in the first place (i.e. waiting to cross a busy road for safety reasons). Next, exaggerated or emotional fear deals with an individual "recalling past fears or occurrences and injecting them into a current situation." This type of fear is particularly relevant to conflict. Emotional fear affects the way people handle conflictual situations.[2]

A police SWAT team executed a raid on a home expecting to find a large cache of drugs. All members of the raiding team were dressed in full combat gear including bulletproof vests, guns drawn, and wearing military battle dress uniforms (BDUs). A female officer, dressed and equipped the same, accompanied the raiders in the event that any children were in the house. Her duty was to take custody of the children and to reassure them. When she entered the room where the children were put, she found an 8-year-old girl and a 6-year-old boy. The girl stepped in front of her brother to protect him and took a protective/defensive stance. The female officer saw the children and began to reassure them they had nothing to fear and that everything was going to be ok.

Twenty years later, the officer met the young girl, now an adult. The girl related that the past years had been very difficult because she suffered so much deep trauma from fear that night. She and her brother suffered nightmares of that night for several years. The boy slept in the same bed with his sister until he was 11.[3]

Summary

Everyone has his or her own concept of reality. Without delving into symptoms of schizophrenia wherein a victim sees and hears people and things that don't exist, we can say if it is not in our consciousness it does not exist, at least not in our current reality. Fear is often a result of unconscious arousal to a perceived threat. Quickly the brain assesses the arousal and determines if the threat is real. The quote, "If the brain doesn't know it, the eyes can't see it" is true to the extent the object has not been cleared by the conscious brain.

Practicing games such as the scenario building of "What if …" allows the eyes to see what the brain has brought into reality because it has been experienced before. Scenario building of "what if" events allows a person to decipher the reality of a real threat over an imagined one.

Review Questions

1. Can your brain "see" something before your eyes do? Can it be called instinct or intuition? Explain the difference between instinct and intuition.
2. If something is not in your conscious view, does it exist in your reality?
3. Discuss the changes in the brain upon experiencing a startle or sudden threat to your safety.
4. Make up a "What if …" scenario and give a friend one minute to solve it. Exchange roles.
5. How does witnessing a traumatic event, such as a fearful experience, affect perception of an event in the future?

Notes

1 Merriam-Webster Online www.webster.com.
2 Barker, Phil; "What is Fear?" www.beyondintractability.org/essay/fear/. July 2003; Retrieved July 9, 2019.
3 Balko, Radley; *Rise of the Warrior Cop: The Militarization of America's Police Forces*; Public Affairs Books, NY, A member the Perseus Book Group 2014.

17

PTSD IN LAW ENFORCEMENT AND COUNTER VIEWPOINTS OF POLICING

Even Police Cry

Post-traumatic stress disorder (PTSD) is loosely defined as "a reaction to trauma as it relates to life-taking or threatening circumstances." The symptoms are not always readily observed and, in fact, may take years before surfacing. No one, especially law enforcement officers, is immune.

> People of all ages can have post-traumatic stress disorder. However, some factors may make you more likely to develop PTSD after a traumatic event, such as:
>
> - Experiencing intense or long-lasting trauma.
> - Having experienced other trauma earlier in life, including childhood abuse or neglect.
> - Having a job that increases your risk of being exposed to traumatic events, such as military personnel and first responders.
> - Having other mental health problems, such as anxiety or depression.
> - Lacking a good support system of family and friends.
> - Having biological (blood) relatives with mental health problems, including PTSD or depression.
>
> **Kinds of traumatic events**
>
> The most common events leading to the development of PTSD include:
>
> - Combat exposure.
> - Childhood neglect and physical abuse.
> - Sexual assault.
> - Physical attack.
> - Being threatened with a weapon.
>
> Many other traumatic events also can lead to PTSD, such as fire, natural disaster, mugging, robbery, car accident, plane crash, torture, kidnapping, life-threatening medical diagnosis, terrorist attack, and other extreme or life-threatening events.[1]

> Even in times of trauma, we try to maintain a sense of normality until we no longer can. That, my friends, is called surviving. Not healing. We never become whole again ... we are survivors. If you are here today ... you are a survivor. But those of us who have made it thru hell and are still standing? We bare a different name: warriors.
>
> *Lori Goodwin*

According to the *DSM-5* (*The Diagnostic and Statistical Manual of Mental Disorders, Fifth Edition*—2013) from the American Psychiatric Association, the following text summarizes the diagnostic criteria for PTSD:

Criterion A (one required): The person was exposed to: death, threatened death, actual or threatened serious injury, or actual or threatened sexual violence, in the following way(s):

- Direct exposure
- Witnessing the trauma
- Learning that a relative or close friend was exposed to a trauma
- Indirect exposure to aversive details of the trauma, usually in the course of professional duties (e.g., first responders, medics)

Criterion B (one required):
The traumatic event is persistently re-experienced, in the following way(s):

- Intrusive thoughts
- Nightmares
- Flashbacks
- Emotional distress after exposure to traumatic reminders
- Physical reactivity after exposure to traumatic reminders

Criterion C (one required):
Avoidance of trauma-related stimuli after the trauma, in the following way(s):

- Trauma-related thoughts or feelings
- Trauma-related reminders

Criterion D (two required):
Negative thoughts or feelings that began or worsened after the trauma, in the following way(s):

- Inability to recall key features of the trauma
- Overly negative thoughts and assumptions about oneself or the world
- Exaggerated blame of self or others for causing the trauma
- Negative affect
- Decreased interest in activities
- Feeling isolated
- Difficulty experiencing positive affect

Criterion E (two required):
Trauma-related arousal and reactivity that began or worsened after the trauma, in the following way(s):

- Irritability or aggression
- Risky or destructive behavior

- Hypervigilance
- Heightened startle reaction
- Difficulty concentrating
- Difficulty sleeping

Criterion F (required):
Symptoms last for more than 1 month.

Criterion G (required):
Symptoms create distress or functional impairment (e.g., social, occupational).

Criterion H (required):
Symptoms are not due to medication, substance use, or other illness.

Two specifications:

- **Dissociative Specification**. In addition to meeting criteria for diagnosis, an individual experiences high levels of either of the following in reaction to trauma-related stimuli:
  - Depersonalization. Experience of being an outside observer of or detached from oneself (e.g., feeling as if "this is not happening to me" or one were in a dream).
  - Derealization. Experience of unreality, distance, or distortion (e.g., "things are not real").
- **Delayed Specification**. Full diagnostic criteria are not met until at least six months after the trauma(s), although onset of symptoms may occur immediately.[2]

The following was written by a long-time police officer whose career came to an abrupt, unexpected ending because of circumstances beyond her control. She became a victim of post-traumatic stress disorder (PTSD). In the past, the term was known as "burnout" or "combat fatigue." It occurs when the victim is exposed to a high tension life or death threat. It may not be the sufferer of PTSD whose life has been in danger, but the exposure to the threat to someone else's life is just as psychologically damaging.

WHAT MOST PEOPLE, INCLUDING COPS, DON'T UNDERSTAND

I've been off work now for 4 months and 11 days. Since that last radio call that had me dealing with a suicidal teenager on the Colorado Street Bridge. Why they don't put safety nets under that bridge I'll never get. Screw the "aesthetics" of the bridge. It's not called "Suicide Bridge" for nothing. And every cop that has ever been up there talking to someone who inevitably jumps to their death knows exactly how fucked up a feeling it is to witness that. They know the helplessness, the inability to "save" a life, the guilt afterward of what you should or shouldn't have said to that person. It's a real mind bender.

So after that incident, I mentally checked out. Fell apart is more like it. To answer your question, no … the teen did NOT jump. He came back over after 4½ hours of talking to us.

3½ hours as me and one other guy as the primary. I'm pretty sure I was pouring my heart out to this kid begging him not to jump, crying intermittently, telling jokes, trying to reason with him, whatever I had to do to keep him on the ledge … and not leaning out towards his death. But the next day at work I was numb. I was empty. I sat in a parking lot and cried for 90 minutes after roll call wondering what the next life or death call I'd get would be. And I realized … I CAN'T BE HERE. So I left. I called the shrink. I saw the shrink. I saw another shrink and then the useless city doctor who referred me to yet another shrink … who ultimately diagnosed me with PTSD and placed me off work.

Therapy has continued for the past 4 months and I've have good and bad days. I've travelled, spent time with friends and family, and had times where I truly forgot about the job and was able to relax and feel like my old happy go lucky self. But still there were the persistent nightmares of people dying around me—and me trying to help them. Still there has been insomnia and days like today where I just CAN'T seem to get out of bed—much less leave the house. The doctor says it's all normal and my brain is "healing" from the cumulative nature of everything I've seen. She said I'm "avoiding" life because it's safer for me. And she is absolutely right. No one is dying or getting run over by cars if I just stay in my living room or on my patio. And I still have this stupid idea that when I DO go out with friends … I will be able to protect them from anything bad that might happen. The reality is, I've seen enough to know … you just can't protect or save everyone. So I stay here in the bubble … where I don't have to be a hero, I just have to be.

I was feeling better a few weeks ago, until during a therapy session with the doctor it finally came out that I can think of no compelling reason to go back to the job. Not one. Not even the money. I simply don't feel that I have anything left to give to law enforcement. I've given so much over the years … that no one really sees, but that I feel … every day. So many dead children, so many ruined children, the suicides, the people who are just lost … I carry so much of them with me, it's unhealthy. I know that. But like the shrink says, that's what made me a great cop—because I truly cared. Well, I can't care anymore. Not for strangers, not like that, not in uniform, and certainly not in patrol where you are at the whim of every radio call. Sure I could handle any call I was given, but it would ultimately lead me back into depression, drinking, and trying to "avoid" life.

So here I am, two weeks after both my therapists and I have come to the conclusion that I simply can not expose myself to the kinds of traumatic incidents that law enforcement has no control over. I'm too "raw" and "fragile" after 21 years spent largely in patrol and working child pornography. I get it. And I know they are right. But with the relief of accepting the fact that I will file for a medical retirement, came the unexpected and overwhelming emptiness and depression I'm feeling over losing a career that I loved so deeply. I have been an officer nearly half my life … and I truly loved my career. I used to tell people it was the best job in the world, but I don't say that anymore. Which saddens me. I don't know how my partners do it … maintain their sanity with all that we see. My hats off to all of them. It's a thankless job and we deal with the absolute worst of humanity, day in and day out. For now, I'm still working on me and getting back to the happy go lucky Micky I used to be. She was awesome! And I pray that one day in the future … she will return.

To my fellow LEO's (law enforcement officers) and Veterans out there—keep your heads up and keep your partners in your prayers. One never knows what they might be internalizing or how difficult a time they may be having at home, in the field, or just in life in general. Don't be afraid to reach out and be a friend … sometimes just a call or a hug might make all the difference in the world.

Thank you all for your support and friendship—it has been humbling.

A few months later, the same (now retired) officer posted a follow up to her story:

Good afternoon friends, fellow LEO's and proud Patriots! I know it's been forever since I've logged on to my police Facebook page, but life has been ever-changing and I'm just now ready to update everyone as to what's new and what's in store for my future …

As many of you know, I've been off work since May 2016—diagnosed with PTSD due to the cumulative effects of years of working the streets and seeing far too much death and destruction for any one person (let alone a mother) to process.

After months of therapy, I finally came to the realization that 22 years in Law Enforcement was indeed enough. It was a tough decision to make, and not one that I took lightly—as I have always believed that being a Police Officer is the greatest, most rewarding job in the world. However, it comes with many sacrifices. Sacrifices endured by family, friends, coworkers and most of all by one's self.

After so many years of working patrol and ICAC, I was no longer able to maintain a healthy and happy outlook regarding my ability to "make a difference." I changed. The job itself has changed very little over the years ... even with the current anti-police rhetoric that is so popular. But for me, I changed.

I became increasingly irritable, snappy, isolated, hyper vigilant and worst of all—depressed. All attributes that I never wanted to associate with myself or bring to the workplace that I loved so much. Realizing that those changes within me were due largely to the traumatic incidents I witnessed at work—but refused to deal with—was a big turning point for me. There is absolutely nothing natural about what a police officer goes through and sees on a daily basis, and year after year of being witness to the worst that life has to offer definitely takes its toll on one's psyche.

Therefore, I made the decision to accept a disability retirement, and to that effect I filed for my retirement in March. The entire process should be completed sometime this summer and I will officially become a "Retired Police Officer"—I still can't believe it some days. However, I am at peace with my decision and know it was the right one to make for myself, my family and The Police Department—who I still hold in the highest regard. It truly is a great organization, filled with terrific cops and Supervisors—who put their life on the line every day in the pursuit of justice and the greater good of society.

What is next for me you wonder? Well, I'm taking the rest of 2017 off to travel and do some adventuring. Travel has long been my passion and I love seeing new places and experiencing new cultures. A move to Tennessee is undoubtedly in my future (perhaps 2018), along with several business ventures that would allow me the freedom to work for myself and make my own schedule. One thing is for certain, I am leaving my career in Law Enforcement feeling incredibly BLESSED, GRATEFUL, INSPIRED, and HONORED to have been a member of the most amazing brotherhood and sisterhood in the world.

Carry on Sheepdogs! I may not be in a patrol car or a uniform anymore ... but I will ALWAYS have your backs!

Source: Shannon Reece, Pasadena, California Police Officer—Retired.

Suicide by Cop

Suicide by cop is a colloquial term used to describe a suicidal incident whereby the suicidal subject engages in a consciously, life-threatening behavior to the degree that it compels a police officer to respond with deadly force.

Jumping off a bridge or walking off a tall building, ingesting pills, carbon monoxide from a car engine, cutting the wrists, hanging, and other more dramatic methods have been used to self-inflict death. A growing phenomena, estimated to be approximately 10–11 percent of suicides, is what has come to be termed "suicide by cop" (SBC).

> The hotly pursued suspect on a freeway or some other type of person coming to the attention of the police suddenly emerges with a gun in his hand, fires off a quick couple shots. Perhaps the rounds are only blank, or he intentionally shoots randomly expecting to miss. Possibly the gun is only a toy. Maybe he is only holding a cell phone that in the darkness is mistaken for a gun. Whatever his actions, his real intent is to draw return fire from the police, knowing the police shoot to kill. His real intention is to commit suicide but does not want to pull the trigger himself. "Going out in a blaze of glory," he fancies the nightly news telling about the "gunfight" police were engaged in, giving them no alternative but to kill him. Upon further review and investigation it is learned the dead person was depressed, was failing in business or had some other motivation for taking his life but chose to be killed by the police.[3]

A Police Officer's View of "Suicide by Cop" and Police Shootings[4]

Suicide by cop is an extremely difficult topic and significant issue that American law enforcement officers deal with all too frequently. It is hard to truly understand the motive of a person who would choose to voluntarily have their life taken by a police officer. Some would consider this an extremely cowardly and selfish act on behalf of the perpetrator, others might consider it noble. Regardless of motive, all parties involved are significantly impacted. The individual is killed, the family of the victim is usually left wondering why, and the officer or officers that pulled the trigger are negatively affected. Nothing from this type of incident results in a positive/pleasurable outcome. These incidents can have life-long psychological, emotional, and physical implications to surviving parties.

The decision by law enforcement to discharge their firearm at an individual could be viewed differently depending upon the ethical theory applied. For example, (Immanual) Kant would argue in his practical imperative that a person could never be used as a means to an end and that police could never kill someone to save the life of another. Most would agree that this is an extremely unreasonable theory and not applicable in the real world. Police officers must be given some authority to protect the public. Today, law enforcement can legally discharge their firearm to stop a threat when the officer believes their life or the life of another is in imminent danger of serious bodily injury or death. The courts acknowledge that officer involved shootings are extremely dynamic and typically only last seconds. Therefore, when an officer is put in a life-threatening situation, they must process information quickly and make a split-second decision to act accordingly. Many times, discharging a firearm is merely a split-second reaction to an officer's training and it is impossible to access the reason why a person is presenting a threat to the officer.

However, it is acknowledged that not all officer involved shootings are as dynamic and unfold as quickly. In these situations, officers have a moral and legal obligation to only use deadly force as a last resort and in self-defense and defense of others. Oftentimes, time, distance, cover, effective communication, and appropriate resources can provide police with some insight into why a person may be acting a particular way. This is never a guarantee though and, regardless, if a life-endangering threat is presented and the officer truly believes this, it would be unreasonable for the officer not to act to save his or her life or the life of another. Officers must always try to utilize all available resources and tactics first if given a reasonable amount of time to do so before they result to deadly force.

Losing Faith in Police—A Concerned Citizen

Here in its entirety is an open public letter written by a concerned citizen who has lost faith in the police and explains his reasoning.[5]

Dear Cop,

A growing number of people distrust you, resent you, fear you, even hate you. Some even wish you harm, and would be glad if they heard you had been killed. Now, you can tell yourself that it's only nasty criminal scum who would ever think that way. You can concoct in your head all sorts of excuses for why so many people now view you as the primary threat to peace, freedom and security. Or you can take the advice of Sun Tzu, and "know your enemy."

If you want to know why those "anti-police" folk think what they think, I can tell you. Because I am one of them. I have little to no respect for those who wear uniforms and badges. I *expect* police to be power-happy, dangerous, irrational and immoral. This was not always the case. I once viewed police as brave and righteous protectors and defenders. I had F.O.P. and Sheriff's Department stickers on my car. I was a proud and vocal advocate of "justice, law and order." (Only later did I realize that "law" is usually the arch enemy of both justice and order.) If you want to know why I no longer trust or respect those who wear uniforms and badges, here are the five main reasons:

Reason #1: "Just Following Orders"

"Enforcing the law" is not automatically legitimate or moral. The most vicious tyrannies in history (under Mao, Stalin, Hitler, etc.) legalized their oppressions, which were then committed "law enforcers." Historically speaking, far more injustice and murder has been committed in the name of "law" than has been committed in spite of the law. And those who directly perpetrated evil have often used the excuse that they were just doing as they were told, just following orders, just doing their job.

The majority of U.S. "law enforcers" are no different, enforcing whatever decrees the politicians may enact, no matter how arbitrary or ridiculous they might be, and then trying to excuse their behavior using the classic cop-out, "Hey, I don't make the law, I just enforce it." You may have even used that one yourself. Just be aware that when you say that, you are essentially admitting that whatever the politicians tell you to inflict upon your fellow man, you will, without question. And that makes you the primary threat to humanity.

Some argue that cops are noble and virtuous, except for the occasional "bad apple." That is false. In a situation where the "law" is immoral and oppressive—as is the case in the U.S. today—any cop who refuses to enforce unjust "laws" will be fired, leaving only those who are willing to inflict injustice upon others whenever the politicians tell them to. The instances of cops refusing to enforce unjust "laws" are extremely rare. The vast majority continue to commit evil while accepting no responsibility for their actions, and even lashing out at any who condemn them for the evil they personally perpetrate. Whether or not you inflict oppression "by the book" or not is irrelevant. Carrying out legalized, officially sanctioned violent aggression still makes you the bad guy.

Reason #2: Legalized Piracy

Despite the silly rhetoric about "protect and serve," the primary job of most law enforcers is to look for excuses to take people's money. Most police are nothing more than glorified tax collectors, always on the lookout—often as a result of "quotas"—for whatever trivial violation or infraction might justify a citation or an arrest, usually for victimless "crimes." Make no mistake: whenever you rob ("fine") or kidnap ("arrest") someone who didn't harm, endanger, or threaten anyone else, that makes you the aggressor, the thug, the bad guy. The fact that politicians gave their official blessing for you to be a bully and a crook doesn't make it okay. In some cases, the legalized piracy is extremely blatant, such as under the "civil asset forfeiture" laws, where police commit armed robbery, stealing cash and other property based on the mere accusation—without proof or trial—that the property was somehow related to criminal activity.

Reason #3: Prone to Violence

The epidemic of police brutality is not merely bad luck. Teaching someone that he has "authority" over others and training him to try to gain control over every situation will naturally cause someone to become an arrogant, abusive bully. If you truly believe that wearing a badge and a uniform gives you special rights, gives you the right to detain and interrogate people, boss them around and control them, then you are automatically going to treat others as your inferiors, and you will lash out at any who don't bow to your supposed "authority." And if you don't truly believe that the badge and uniform gives you special rights, you will either quit or be fired.

"Abuse of authority" is redundant. History shows one heinous example after another of what people become when they are given power over others. It should be clear to anyone who understands psychology and human nature why police have much higher rates of domestic abuse. Your job trained you to be a violent control freak. If and when your violence changes from "legal" violence to "illegal" violence, why should anyone be surprised? You are a professional thug.

Reason #4: Gang Mentality

When a criminal cop abuses, assaults, or even kills an innocent person, the vast majority of other cops can be counted on to take the side of the badge-wearing perpetrator. Police unions and departments are constantly making excuses, even for outright murder, always insisting that whatever anyone with a badge did must have been justified. "It's a dangerous, stressful job, you don't understand what it's like, and you don't know the whole story, and he feared for his life." The same old broken record plays over and over again, as one example of police brutality after another comes to light. The "bad apple" argument rings hollow when all the other "apples" defend and protect their fellow gang members, no matter how egregious, illegal, or immoral their behavior. Where are examples of these supposed "good cops" arresting the bad ones, or at least doing something to try to stop police abuse? Instead, one video after another shows that when a thug with a badge is committing evil, all the other cops in the area will, at best, passively watch the abuse like spineless cowards, and at worst, will join in like sadistic thugs.

In short, "law enforcers" make up the most dishonest, violent street gang in the country, loyal only to their fellow gang members, showing little or no concern for anyone else. It is sickening how quickly and easily badge-wearers routinely endanger, threaten, assault and even kill others, with little or no provocation. Then they try to cover their paranoia, sadism and criminal negligence with excuses about how dangerous the job is, which is statistically untrue anyway.

When an average person is killed, the police fill out a report about it. When one of their fellow gang members dies, they will invest all the time, resources and man-power they have at their disposal, often endangering, harassing, terrorizing, and attacking innocents in the process, as happened with the Eric Frein and Christopher Dorner cases. It is obvious which lives matter to police, and which don't.

Reason #5: Criminals with Badges

The gang mentality, and the disregard for the safety and freedom of everyone else, naturally leads to police becoming criminals themselves, lying under oath, planting evidence, selling drugs, running extortion rackets, and so on. Police are only human beings, but human beings who are trained to control and extort others, trained to view and treat others with disdain, and trained to believe that they are so important and righteous that basic morality doesn't apply to them. Add to that the fact that police "take care of their own," and rarely investigate or prosecute their fellow officers for criminal conduct, and it is obvious why so many cops become the criminals they pretend to be protecting society from. For example, the Philadelphia narcotics squad was just outed as a bunch of drug-dealing extortionist thugs. And the gang members in blue know that they can almost always count on their fellow gang members to look the other way when they break the law themselves. When one officer does have the fortitude and courage to speak or act out against police misconduct, invariably all the other badge-wearers will gang up on the "traitor," for having chosen truth and justice over loyalty to the gang.

Where This Road Leads

A look back at history strongly suggests what the outcome of this situation will be. Police will become more brazen, more criminal, more violent, and no amount of protests or political action will do a thing to stop it. Every indication is that the power-happy thuggery of those who imagine themselves to be "authority" will naturally get worse and worse, until the people have no choice but to violently defend themselves against the thugs in blue. And that will only make the cops less restrained, more paranoid, more trigger-happy, more criminal, and more violent.

Making excuses for your thuggery is not going to make people trust you or respect you more. If you actually want the respect of the people, if you actually want to be a force for truth and justice, then start publicly condemning police misconduct and abuse. Start arresting the criminals who wear badges. Start treating other people as your equals, instead of as your subjects. And if you don't, then don't be surprised when the people resent you, distrust you, fear you, and hate you. If you are content to think and act like just another street gang (albeit a very well funded and well armed street gang), becoming ever more lawless and violent, don't be surprised if one day one of your intended victims takes it upon himself to enforce a little justice of his own. You will have earned it.

In a recorded video, one police officer took a counter stance in his appraisal of policing and police attitude:

> A courageous Georgia police officer, identified as Billy Ray Fields, has taken a strong stance against the rampant police abuse of authority. In a viral video, originally posted on *Mediatakeout*, Fields states that he supports the police accountability movement as a means of ridding departments of the bad cops.[6]
>
> "When I became a cop, I took an oath knowing what I would face in these streets .... This young generation they're our future and some of them don't have a clue nor a plan. So it's not only my job to uphold the law, but to show, guide and lead them in the right direction ... I'm the same guy with the badge on as well as off. I give the dope boys and gang members the same amount of respect and courtesy as I would a judge or a lawyer .... Because at the end of the day we've all done bad at one time or another. It's not my job to judge how you choose to live your life, as we are all one nation under god .... My words are real and I stand behind what's right. So to the actual good cops out there I salute ya'll, stand up, cause this too shall pass. But the bad cops out there, man go cut yourself and go jump in the ocean for me. My people, continue to record. I'm Billy Ray and I'm 10–7 out."

In a survey created by the author, one police respondent noted,

It's not the police who need to be retrained, it's the public. We have grown into a mouthy, cell phone wielding, vulgar, uncivil society with no personal responsibility and the attitude of "it's the other person's fault," "you owe me." A society where children grow up with no boundaries or knowledge or concern for civil society and personal responsibility.

When an officer says "Put your hands up," then put your hands up! Don't reach for something in your pocket, your lap, your seat. There's plenty of reason for a police officer to feel threatened, there have been multiple assaults and ambushes on police officers lately. Comply with requests from the officer, have your day in court. Don't mouth off, or fight, or refuse to comply … that escalates the situation.

Police officers are our sons and daughters, fathers and mothers, brothers and sisters. They're black, white, brown, all colors, all ethnicities, all faiths, male and female, they are us. They see the worst side of humanity … the raped children, the bloody mangled bodies of traffic victims, the bruised and battered victims of domestic violence, homicide victims, body parts … day after day. They work holidays while we have festive meals with our families.

They miss school events with their kids, birthdays, anniversaries, all those special occasions that we take for granted. They work in all types of weather, under dangerous conditions, for relatively low pay. They have extensive training, but they are human. When there are numerous attacks on them, they become hyper vigilant for a reason, they have become targets. When a police officer encounters any person … any person, whether at a traffic stop, a street confrontation, an arrest, whatever … that situation has the potential to become life threatening.

You, Mr & Mrs/Miss Civilian, also have the responsibility of keeping the situation from getting out of control. Many law enforcement officers are Veterans. They've been in service to this nation most of their lives, whether on the battlefield or protecting us here at home. They are the only thing that stands between us and anarchy in the streets. If you want to protect your child, teach them respect.

Summary

Post-traumatic stress disorder (PTSD) is a medically and psychologically recognized disorder formerly named "combat fatigue" or "burnout." PTSD is a consequence of exposure to a life-threatening event. It may not be a singular event but an accumulation of events over a number of years. Police officers have a high percentile chance of being victimized by PTSD simply by the definition of their occupation. Years and years of seeing the worst in humanity and escaping life-threatening circumstances can make even the strongest and bravest feel the pain and suffering of PTSD.

A society needs police officers to maintain law and order to keep society from becoming a terrible dystopian and violent culture ruled by the strongest and the most violent. No matter, there will always be those individuals and groups who have no respect or feelings other than disparagement and contempt.

Review Questions

1. Describe the symptoms and consequences of PTSD.
2. How is PTSD developed?
3. Is the writer of the "Dear Cop" letter correct in his assertions of police brutality and corruption?
4. Do you believe the officer survey respondent correctly identified contemporary society and what it means to be a police officer?

Notes

1 www.mayoclinic.org/diseases-conditions/post-traumatic-stress-disorder/basics/risk-factors/con-20022540. Retrieved January 27, 2017.
2 www.brainline.org/article/dsm-5-criteria-ptsd. Retrieved October 21, 2019.
3 June, Dale L. and Rebecca Stincelli in *What They Didn't Teach at the Academy*; Edited by Dale L. June; Taylor-Francis, CRC Publishing, Boca Raton, FL 2015- Rebecca Stincelli is a recognized expert in the area of Suicide By Cop. She began researching "suicide by cop" in 1986 while working with the deputies of the Sacramento County (CA) Sheriff's Department as a crisis interventionist and post-trauma liaison for victims of violent crimes. During the next decade, Rebecca became a law enforcement instructor where she was given the opportunity to present the perspectives of both the loved ones of the decedent and the officers involved.
4 This particular section was written by a National University Criminal Justice student and veteran law enforcement officer.
5 Rose, Larken; *5 Reasons Law Enforcement in America Creates Fear and Resentment Instead of "Safety"*; The Free Thought Project; July 5, 2015. http://thefreethoughtproject.com/5-reasons-law-enforcement-america-creates-fear-resentment-safety/#qY2OJzCfWvtUklc7.99. Retrieved September 21, 2019. Used with permission.
6 Syrmopoulos, Jay; *This is What a "Good Cop" Looks Like, Here's the Video All Cops Should be Making*, June 16, 2015. http://thefreethoughtproject.com/courageous-cop-crosses-thin-blue-line-supports-police-accountability-movement-tells-people-continue-record/#hmeITudvtBmmQirG.99. Retrieved April 16, 2017.

18

FINAL CONCLUDING THOUGHTS OF SOCIAL PSYCHOLOGY AND FEAR

In Austin, Texas in March 2018, an unknown person used a near invisible "trip wire" to set off explosives, killing two and injuring four over a time period of three weeks. The fear caused by the unknown and the random danger quickly spread throughout the city by news and social media. Cities with a large gang influence know fear from numerous drive-by shootings, never knowing when one will occur or who will be the victim. Our modern day's generation of society has witnessed airplanes being used as giant bombs to bring down tall skyscrapers and smash into the Pentagon; mass shootings and killings in movie theaters and churches, surveillance cameras everywhere, searches and screenings at airports and public buildings, people living in gate-guarded communities and installing intruder alarms in their homes, riots and violence, shootings involving the police and "aggressive suspects." An atmosphere of fear runs close to the surface. A recent ad ran in a local movie theater urging patrons to "report any suspicious circumstance or person." In another episode, movie goers were stricken with fright and began panic-stricken running from the theater when someone opened a back exit door and started a noisy leaf blower. They reported they thought someone was attacking with a chainsaw.

To repeat a previously used cliché, "The eye can't see what the brain doesn't know." How does the brain see and recognize danger leading to fear?

> New York university professor, Dr. Joseph LeDoux spent decades studying fear in the brain. He described two pathways of differing lengths that fear signals follow within the brain. His experiments suggest that our sensory organs pass information to the thalamus, where signals split and trace two separate paths on their way to the amygdala, the brain's threat center. Following the shorter path, one signal manages to sound the fear alarm before we're aware of the situation. The other reaches the sensory cortex a fraction of a second later, providing a much clearer picture of the potential threat. The second signal can reinforce the fear response or declare a false alarm.[1]

In a speech to the Muslim community, former U.S. Attorney General, Loretta Lynch said, "When we are ruled by fear, we are not making ourselves safe."[2] This statement can be taken in several contexts. It can be taken to mean that when danger abounds, we need to step up and not be frozen by fear or take flight. We need to confront fear—the fight aspect of the adrenalin flood. Among all human needs, the need for security is just above the physiological needs of oxygen, water, food, and sex. Foremost among the security needs is the need for physical safety. From the first moment of birth to the last expelled breath, humans need the reassurance that they are safe from any hazards that would injure or kill them. Among the ocean of fears for their safety and security is the fear of entanglement in legal matters and being a victim of a crime.

One burglary victim, a college student, expressed the traumatic and emotional impact the crime had on her.

> Our television was missing and things were on the floor, important papers were left scattered on the floor. Our refrigerator was left open with most of our food missing, but what I will always remember is my parents because all of their room drawers were open, clothes were on the floor, the bed was unmade, and what hurt me the most is that our gold jewelry was missing. I couldn't do anything but be frightened and cry. I couldn't believe someone had done all of this damage to our home. I can remember asking my parents, "Who would do this?" "Why did they do this?" My home doesn't feel like home any more because now I am scared to be there wondering if they would come back again.[3]

Unfortunately, beside the fear of being a crime victim is the thought of being the target of a police action involving something as simple as a traffic infraction for failing to signal when changing lanes on a highway. When the driver of a vehicle sees the red and blue lights of a police officer following him and indicating for him to "pull over," the driver begins a series of physiological changes, ranging from anxiety and dread to fear and anger, contingent on the circumstances.

A marked police car stops in a residential street and two uniformed officers get out and begin walking toward a specific house. Neighbors look out through their curtained windows to see what it is all about, glad the officers aren't coming to their door. If the resident of that particular house hasn't called the police for some type of assistance, he will be filled with anxiety and dread, expecting some kind of problem and the officers to be unfriendly. Of course if he has called the police, he will experience relief, knowing that his problem will be taken care of by the friendly officers.

> Black people fear and avoid the police and will tolerate criminal behavior sometimes just to avoid involving police whom they fear more than anything one can perpetrate on another. As they have witnessed calling the police for help in a distressful situation only to have the police turn on the one who called or go too far when they only sought a peaceful intervention. It's nothing like in white neighborhoods. Completely different behaviors for the reason of fear. Where police constitute the biggest fear. Not because the people are criminals but because of the criminal acts the police have perpetrated on those people.[4]

Some people become more anxious and fearful of crime than others because of certain experiences in their lives, including how they were raised and the environment they were raised in. Growing up without knowing crime and violence and not knowing the fear of it is enough to make one fear what is out there. Afraid of the unknown, afraid of that "what if something happens?" Some people experience horrendous traumatic instances in their lives, which leave them apprehensive and afraid. A person experiencing being a victim of a home burglary, even while away from the residence, will become apprehensive every time an unusual noise is heard or imagined. They become easily startled, believing the burglar is returning to perhaps steal more items or to hurt them. They feel violated their private space has been invaded by a person or persons unknown. The feeling of their vulnerability will linger for an extended period of time, perhaps even years.

A police officer was awakened one late night when he heard the sound of someone breaking into his house. He immediately rose from his bed and picked up his hand gun. Moving quietly, he checked the bedrooms and saw that his children were asleep in their beds. Continuing to the kitchen area he saw the dark figure of a man trying to open a window. The officer quickly opened the window and stuck his gun into the face of the intruder, ordering him to stay as he was. The man complied and was held at gunpoint until deputy sheriffs arrived and took him into custody. It was later learned he had stumbled drunkenly from a party at a nearby neighbor's and believed he was at his own house. Though the incident was merely the mistaken action of a drunken man, his activity led to a critical moment that could have fostered his death. For months after the incident, the officer and his family became keenly aware of any noise that seemed unusual. The wife and children became anxious about the possibility of being victimized and this caused them to install a home security system.

There are many factors involved when it comes to fear but one way to overcome it is to have personal experience with crime and violence and prepare mentally for what might happen instead of living in fear that you might be the next victim. Fear in society and who is fearful have a lot to do with the area one is raised in. The individual themselves and their gender are also a factor. Further, what type of household one grows up in and what type of social media one is exposed to are contributing factors of fear in today's society.

Inner city inhabitants see more violence and are prone to thinking that same level of violence exists everywhere. Seeing shootings, violence, crime, and even death becomes normal so that, instead of being afraid, one embraces it as a way of life. Growing up in an environment in the suburbs with little crime and violence, with the knowledge that it happens "somewhere else," and so not knowing the fear of it, is enough to make one anxious when leaving one's safe home. Afraid of the unknown, afraid of "what if something happens?" "Lock the doors or someone will break in," "don't go down that street because there are bad men there," "don't walk by yourself or someone will take you," "don't talk to strangers," "don't be outside in the dark because it's dangerous" etc. ... All these rules of what is bad that we learned growing up affect what makes us fearful in society. Social media also portrays society as corrupt, dangerous, and frightful most of the time, although it may not be intentional but it can be taken that way.

Seeing crime, violence, and death on television is chilling enough to make people fearful of what could occur to them once they step out into society. Having a constant flow of violence on the news of white police shooting black teenagers terrifies and instills fear into the minds of the black community that they could be next. From these actions rise racial tensions between white and black communities and now there is fear on both sides that they or their family and friends could be targeted.

One child grows up in the "racially mixed neighborhoods" of East Los Angeles and another grows up in the friendliest neighborhood of Calabasas. Both of these children are going to more than likely have different perspective on what fear is. Growing up in East Los Angeles one would likely be exposed to seeing drug deals go down, robberies in the local groceries stores, drive by shootings, children being led into gang life and choosing sides between friends, witnessing a shooting, seeing a friend killed, and constantly hearing the ambulance, firefighter, and police sirens day and night.

Growing up in a safe neighborhood like Calabasas, California, one may have grown up leaving toys and bikes outside overnight, not locking the doors, and not exposed to the brutality of robbery, rape, and murder. But it is normal because the crime rate is low and the fear of anyone committing a robbery or theft is minimal, if at all present. It is safe to walk anywhere without any trouble from anyone. Violence is seen on the nightly news but never experienced or seen first-hand.

Blood, guts, gore, hair, and eyeballs splattered all over the room, street, or highway. We have become desensitized to violence and death; we see it every day on the news, in entertainment (violent games and movies), war (faceless and depersonalized drones killing from the sky), acts of terrorism (by definition violent death), that every instance of mass killings seems to set new records for depravity and body count. On October 1, 2017, a shooter in Las Vegas shot and wounded 500 people and killed 58. What new madman will soon step up to take a new record number of victims?

Guns and Violence

> Since gun owners do not agree with high-profile gun control measures, we must therefore support the murder of children.
>
> *Unknown Facebook contributor*

Another anonymous Facebook contributor added:

> I am a criminal. I don't buy my guns at stores. I don't care about your stupid gun laws. Background checks won't stop me. Keep focusing on the good people. That is exactly what I want.

It's not the instrument, but the musician who makes the music. A trumpet by itself is nothing but a piece of steel. It needs someone to blow it. "Take away the guns!" becomes the cry of the day. Put up walls and barricade our schools, homes, and communities. Turn the entire country into a prison. Take away freedoms, monitor everyone, friends, family, and neighbors spying on each other to report "suspicious behavior."

When I was in the sixth grade and learning proper English, the teacher was explaining the use of auxiliary or "helper" words such as had, have, and has. "When using 'begun' you always need a 'helper' word because a gun always needs a helper to make it work." In cases and times like this, the gun is merely the instrument and the shooter is the "helper." It is not the gun that is sick, but the mind of the person holding it.

Let us not medicate ourselves by thinking guns are the villains; it is the warped mind of the individuals who wish to "cleanse" the world of what they perceive as "demons" or those who seek revenge for some imagined slight, offence, or cause. Desensitization of death by emphasizing it as a resolution to personal problems and as "games" has only made the minds of those "sick individuals," those "lone wolves," those "friendless outcasts" in society who believe they are "Zendor, Ruler of the Universe," plan and execute more and more heinous crimes.

Every time something like a mass shooting occurs, we can hear the clink of another freedom being lost. There is no government conspiracy to remove guns, it is the fanatical "kneejerk" reaction by people whom I can relate to only as "lemmings." It's not the guns but the mentality of people who think of guns as "equalizers" and perhaps don't appreciate the finality of death. Death is never pretty. I have had guns since I was 12 years old, and I used to hunt, but to see a magnificent animal dead at my own hands stopped me from hunting.

What makes people fearful of guns is ignorance; people always fear what they do not understand and there are many Americans that don't understand guns or even individual rights for that matter. What we need in schools is not less talk of guns; it is more training on proper usage and impact that gun usage can have. We need more education in this country that we are severely lacking about guns and gun violence and the actual statistics of gun violence.

A gun is no more than a tool that can be used for recreation much like a large knife can be a kitchen tool to prepare meat. It can also be used to kill or maim a person either in cold blood or in self-defense. People seem more intent on getting rid of guns due to the record number of homicides, mass shootings, and accidental shootings. Every year there are deaths by vehicle yet nobody seems intent on banning those. Most people are anti-gun because they don't understand how to use one. Once understanding occurs they will be less likely to be anti-gun.

Mass Shootings

People will always be afraid of the unknown and what they don't understand. My point is that people who are afraid of cars, for instance, don't know how to operate them and don't understand their capabilities, even though in our society having a vehicle is very normal. People who are anti-gun generally do not know the first thing about a weapon and just know that they are used for killing.

Even in street-gang drive-by killings there are always more casualties than the immediate victim. These include the families of both the perpetrator and the immediate victim, everyone in their circle of friends and relatives, witnesses, and first responders. The ripple or "butterfly" effect is in full play.[5]

Games and other modes of entertainment seem to become bloodier and bloodier as the makers spill more and more red ink to rake in the "green." We cannot "pacify" the world by disarming, but we can make it a better world by learning concepts of peace, love, and harmony and teaching the finality of death. What effect would even one suicide-murdering bomber make in a crowded New York subway, nightclub, or other busy venue? Fear would become the element of society, knowing that suicide bombers have begun operating in the United States. Of course spreading fear is the *raison d'être* (reason for being) of terrorists.

A letter to the editor of the *Los Angeles Times* on November 29, 2015 seems to place fear and security over freedom:

> I do not understand why people are so paranoid about government spying. I don't sell drugs or, espouse violence, so government agents can read all my emails, listen to my phone, look at my health records and put a camera in my bedroom. This fear (of government spying) seems to be ego based, in other words, you are not as important as you think … I for one, like my government, trust it and want it looking for bad guys everywhere.

An attitude of "I'm not doing anything wrong so why should I be afraid of government watching over us?" is an attitude that opens many doors to government over-watch, which should be more frightening than the writer of that letter can even imagine. Many, or most, police officers have expressed concerns about police departments utilizing drones to patrol the skies over neighborhoods and businesses. They are in agreement that a drone deployment is a useful tool to aid in hostage and perhaps kidnapping situations, but are very much against "routine" drone neighborhood watch because of the fear of an invasion of privacy.

I think society in general with overpopulation and frustration of modern stress and violence bred into our minds and souls by "entertainment" and the mindless concern people have toward the reality and stink of death and the feelings of others are perhaps as much to blame. I advocate reaching out to others, every day doing one spontaneous good thing for someone and expecting nothing in return. Sometimes a smile, a handshake can save a life—we never know. We also never know who needs a smile, a handshake. My studies of serial killers, assassins, "mob hitmen," and other "heartless" killers or "sociopaths" have shown me that those individuals have nearly all experienced violence or being "outcasts" in their younger years. Where is the love, kindness, caring, the spirituality of "brotherhood" of man toward our brethren? We must as a society look out for the "other." Not necessarily be a "keeper" but recognizing when someone is about to "go around the bend" and do what we can to help him/her. Sometimes a smile, a handshake …

I have had guns, as I earlier stated, since I was 12. Even my high school had a rifle club. We shot in the school hallways after school was closed for the day (we shot .22 shorts rounds after securing the steel fire doors). I learned gun safety and how to handle a gun with respect. A gun was on my side for most of my adult life. If we disarm the population, who will have guns? … the police and bad guys. As a lifetime law enforcement officer, I dare to ask "Can we always trust the police or bad guys?" Sometimes they are the same person. Can we always 100 percent trust the government who controls the enforcement community? I am not a "government is untrustworthy" conspiracy theorist, but when I see freedoms being neutered in the name of security because of incidents like this I become concerned. I have always spoken that we can lose 100 percent freedom but never gain 100 percent security.

A mass killer kills more than four persons at a time in a onetime event, either operating as a lone sniper (Charles Whitman—University of Texas, Austin, August 1, 1966: 15 victims; high school freshman, Brenda Spencer, January 29, 1979: killing two and wounding seven in the first known modern school shooting) or as an intruder surprisingly spraying bullets around a crowded venue (James Huberty, McDonalds Restaurant, San Ysidro, CA July 19, 1984: 21 victims; Richard Farley, February 16, 1988: 7 victims in a workplace/stalking incident). In the years since there has been a growing number of similar shootings. (Note: these examples are used to illustrate the history of mass killings in the United States since the Whitman case, considered the first of the long list of mass killings, including school shootings.)

Previously mentioned is the Las Vegas shooting at a country music concert on October 1, 2017, when 59 concert goers were killed and over 500 wounded by a shooter using ten guns. Just over a month later, on November 7, a former Marine machine gunner walked into a Thousand Oaks, California bar, crowded with university students and killed 12, including a deputy sheriff responding to the shooting.

The news media rushes to the scene, describing it in the worst and most sensational terms, and "anti-gun" activists and pacifists "come out of the woodwork" like termites.

The difference between a mass killer and a serial killer is that a serial killer kills one person at a time over a stretch of months or years and usually kills with a knife or similar instrument, often strangling the victim. There is nothing more intimate than entering another person's body. Thus the weapon of choice. A serial killer exemplifies this with the "up close and personal" stabbing, often over and over, and perhaps dismembering the body. Some serial killers have killed upward toward, and perhaps beyond, 100 or more.

To avoid anxiety, even panic, in a city when police become aware of the existence of an active serial killer, rapist, or bomber, they will, more often than not, not disclose the information.

A "gangland" killer kills one person at a time, with a few exceptions when more are killed at once, i.e., the Valentine's Day massacre (Chicago, February 14, 1929) when seven persons were killed. The most notorious of "mob" killers is Richard Kuklinski (known as the Ice Man), whose favorite method of killing was with cyanide poison but he utilized various other methods including a chainsaw and sawn off shotgun. He confessed to killing "over one hundred."

In the history of serial killers there has not been one call for the elimination of knives, chainsaws, or cyanide.

An examination of the similarities of mass killers, serial killers, and "mob hitmen" shows one distinctive feature: they were unfeeling about what they did and showed no remorse (psychopathic behavior). Another glaring similarity or commonality is that none of the perpetrators "snapped." Evidence has shown precise planning prior to the event (a serial killer and professional assassin, i.e., "hitman," plans and executes an intricate plan that will enable him to escape—a mass murderer usually plans and stages the event but makes no arrangement for escape and is usually killed or commits suicide at the scene). For the most part, they all are "loners," anti-social personalities (sociopaths), often the product of a broken home, abusive and over-demanding parents, perhaps a harsh father, or overbearing, controlling mother.

An "emotional" killing, such as a workplace violence shooting, a school shooting, or a "spontaneous" killing in "the heat of passion," a "once in a lifetime event," may follow a model such as this:

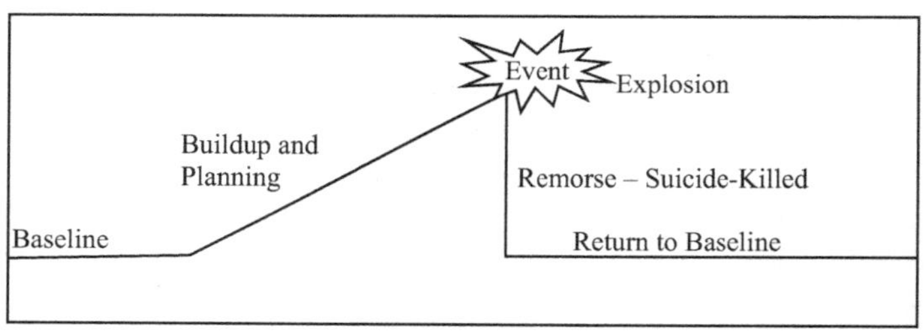

FIGURE 18.1 "Emotional" killing model

It may take years to build up to the explosive event. Frustration, mistreatment by fellow workers and supervisors, alienation, or obsession—as in the case of the mass shooting and killing carried out by Richard Farley at Electromagnetic Systems in Sunnyvale, California on February 17, 1988 who was obsessed with a co-worker, Laura Black.

A serial killer follows a similar pattern of buildup to the killing event but the step to remorse (if there is any) is short lived and, as he continues killing, the steps and time from a return to baseline to the buildup and planning stage become shorter and shorter, until killings become more rapid in sucession. A serial killer can be compared to a drug addict who tries it once, likes it, and continues using until a tolerance is built up, thus needing more and more to arrive at the same feeling, but never again achieving the same "climactic high." He may also kill just to reach back up to baseline and may never reach "normal baseline." The model can be framed like this:

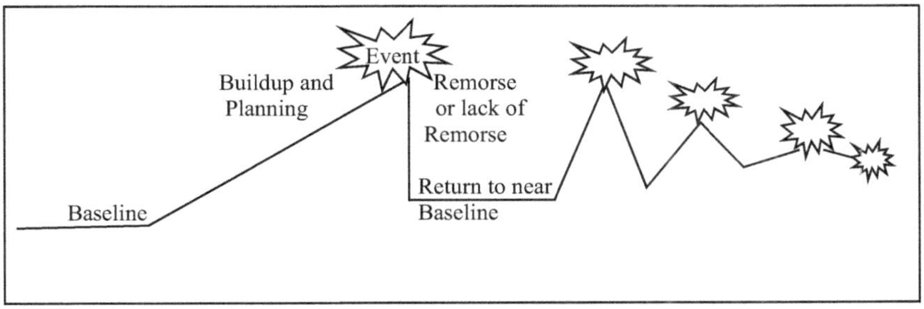

FIGURE 18.2 "Serial killer" killing model

Until he is identified and caught, a serial killer will spread fear and anxiety throughout a city as neighbors wonder and ask, "Who is he? When and where will he strike again?" Knowledge of a serial killer or pedophile wandering the streets is a scary thought in itself but knowledge that one could be in your neighborhood definitely spreads fear through the community. Most people are afraid of death, it is the thing that will come for all of us one day. Feeling that today could be your last day or one of your love ones' last day because of a serial killer trolling your neighborhood would have any normal person in fear, if not for themselves then for their loved ones. The same principles apply to a pedophile, but now it's mainly the fear of someone causing harm to your children, or children in your area; it is a fear of the unknown, fear that something could happen and you would have no control of the situation.

Because anybody could be a potential serial killer or pedophile, many people might share some of the characteristics. Perhaps the grumpy old man down the street likes to be left alone but has been seen waving at little kids because he misses his own, potentially leading people to believe that *he* is the pedophile in the neighborhood they heard about. Or maybe the young nerdy shut-in guy who lives at home with his parents could be the serial killer they heard about on the news. The knowledge just adds to the public's fear, multiplying it unnecessarily.

Fear Moving toward Dystopia; A Loss of Privacy and Individualism—A Look to the Future of Man

American philosopher, Eric Hoffer, presents the theme that:

> Those who would transform a nation or the world cannot do so by breeding and captaining discontent or by demonstrating the reasonableness and desirability of the intended change or by coercing people into a new way of life. They must know how to kindle and fan an extravagant hope. It matters not whether it be hope of a heavenly kingdom, of heaven on earth, of plunder and untold riches of fabulous achievement or world domination.[6]

Characteristic of the truth of human nature is that we think in terms of existence and survival; anything else is counter to that natural instinctive order.

> [Man] must be stripped of his individual identity and distinctiveness. … The most drastic way to achieve this end is by the complete assimilation of the individual into a collective body. The fully assimilated individual does not see himself and others as human beings. When asked who he is, his automatic response is that he is a German, a Russian, a Japanese, a Christian, a Muslim, a member of a certain tribe or family. He has no purpose, worth and destiny apart from his collective body; and as long as that body lives he cannot really die.[7]

Where we are in time, space, and technological advances brings enlightenment to the prophesy of the likes of Arthur C. Clarke and Stanley Kubrick in their classic *2001: A Space Odyssey*. Man's journey begins in the dark mists and swamps of time over a span of several million years. Through evolution he develops a whole new world with the advent of tools and eventually to the artificial intelligence of computers. The HAL 9000 computer sees man as "bored and boring, having to be virtually dead to just get around, humans are merely maintenance men at the end of their evolution. Man's tools [the computer] no longer needs man."[8]

Other prophetic dystopian stories, such as Orwell's *Nineteen Eighty-Four* and movies such as *I, Robot*, and *The Terminator* and *Star Wars* series, give us more than a glimpse of the future of Man as he builds his own evolutionary replacement in the forms of machines, computers, and robots. We wonder, where will the Winston Smiths and Sarah and John Connors or Jyn Erso (of *Rogue One: A Star Wars Story*) or Princess Leia come from? Slowly they are coming in the mold of Edward Snowden and other "revolutionaries" who have the foresight, willingness, and, yes, bravery to challenge governmental control of individual privacy and tracking by machines and supercomputers.

A valid question, especially for those who believe "whistleblowers" like Edward Snowden are traitors and should be tried for treason is: If security means a tradeoff with a loss of privacy, potential incarceration for "thoughts," marching in unison to work or schools while wearing required uniforms, taking away individual choice, and should a revolution be staged against governmental tracking of free individuals, which side would you be on?

Whenever there is a "people's march," "demonstration," or public display of grievances against the government, police respond in a militarized manner or as ordered by those in political power. When asked the question "What side would you be on?" police officers nearly unanimously reply, "I'd support the law" … even if they feel a moral bonding and agreement with the protesters. They would not risk losing their jobs.

The average person's details are in at least 300 databases—from credit bureaus to the NSA. With the advent of social media, the invention of bar codes, drone aircraft, infrared imaging, and the consolidation and increasing power of corporations, it is not remarkable that the whole world, and Americans in particular, are no longer protected from prying eyes.

Everything You Want Is on the Other Side of Fear

> God placed the best things in life on the other side of terror; on the other side of your maximum fear are all the best things in life. Everything you want is on the other side of fear.
>
> *Will Smith*[9]

PERSON IN THE MIRROR, WHAT ARE **YOU** *AFRAID OF?*

Summary

Contemporary society is awash with fear. Fear is taught to the young by parents, fear is spread by rumor, news and social media. The government makes proclamations about terrorists and illegal aliens lurking behind every bush and tree. Security measures are in place in public shopping malls, apartment complexes, government buildings and court houses, churches and schools. Mass shootings and serial killers dominate conversations. Society is becoming a state of paranoia-schizophrenia.

As fear spreads throughout society, cultures change, people become mistrusting of others, the call for "gun control" becomes louder and more guns are sold for personal protection. Militias and neighborhood crime watches are formed and the government becomes more and more "Big Brotherish."

The public turns to the police for protection but, way too often, is met by gunfire by frightened police officers in situations when they may believe their own life is in danger. This all leads to the ingredients and makings of a dystopian future. Fear is living in a state of anxiety that something terrible might, could, possibly, happen. Fear is beyond phobia. It is anticipation of personal threats to one's life. Fear is a predator, sucking life from happiness and well-being. Is it man's future to be in a constant state of fear or can society learn to live beyond fear?

Review Questions

1. Do you agree or disagree that the bravery this country was founded on is no longer a part of the human psyche or spirit?
2. Do you believe you can become a victim anytime, anywhere? Why?
3. Do you believe police over-react to a potentially dangerous situation because of their mindset of believing every situation can be life threatening?
4. Is fear in society being overstated with all the security technology available?
5. Are your feelings of security shaken or influenced by news and social media?
6. Because of the recent upsurge in reported police violence and the public's perceived police use of unnecessary force, is the public's fear and perception justified or is it magnified unrealistically?
7. Do you believe a law-abiding citizen has cause to fear the police, based on the many incidents of police shootings and beatings?

Notes

1 Goosebumps, the Science of Fear; www.fearexhibit.org/brain/wired. Retrieved July 28, 2015.
2 *Los Angeles Times*; "U.S. to Probe Muslim Boy's Arrest," December 5, 2015; p. A 7.
3 Interviewee wishes to remain anonymous, extract used with permission.
4 Dr. Maisha Pesante.
5 The butterfly effect is a term used in chaos theory to describe how small changes to a seemingly unrelated thing or condition can affect large, complex systems. The term comes from the suggestion that the flapping of a butterfly's wings in South America could affect the weather in Texas, meaning that the tiniest influence on one part of a system can have a huge effect on another part.
6 Hoffer, Eric; *The True Believer: Thoughts on the Nature of Mass Movements*; New York; Harper Collins; 1951, p. 9.
7 Ibid. p. 62.
8 See *Kubrick 2001: The Space Odyssey Explained*: www.kubrick2001.com/en/1/index.html. Retrieved July 2, 2019.
9 www.bibliatodo.com/En/christian-news/god-placed-the-best-things-in-life-on-the-other-side-of-fear-says-movie-star-will-smith/. Retrieved March 13, 2019.

INDEX

References to figures are indicated in *italics*. References to endnotes consist of the page number followed by the letter 'n' followed by the number of the note.